Table of Contents

introduction

ARE WE THERE YET? BUILDING AMERICA'S TRANSPORTATION INFRASTRUCTURE NETWORK

By Tom Hill, Chief Executive, Oldcastle Materials, Inc.

Those of us who build America's transportation infrastructure know that our work is largely taken for granted. Few commuters pause on their way to the office to give thanks for the road or rail network they are traveling on. We know that. But we also take pride in knowing that the highways, bridges, rail lines, airports and waterways and ports we build are *used* every day by millions of Americans and are critical to every aspect of our society. We know that without the efficient transportation infrastructure system we build and maintain, American life as we know it would grind to a halt.

If few people stop to acknowledge the day-to-day work of the transportation construction industry, fewer still are aware of the extraordinary role our industry plays in times of national emergency, such as the September 11, 2001, terrorist attacks at the World Trade Center in New York City.

While our nation rightfully offers heartfelt praise to the many heroes of the

rescue effort—particularly police and firefighters who showed such courage in their effort to help others—the tremendous work of the men and women in our transportation construction industry should also be recognized.

As is usually the case, the work of our industry at the site of the World Trade Center disaster has gone on quietly, unnoticed even in the glare of the media spotlight. But it has gone on 24 hours a day, and its contribution has been enormous.

Americans are not likely to know, for example, that **Tully Construction**, of Flushing, N.Y., was at "ground zero" within minutes of the attack, its employees helping to clear debris and search for survivors. Tully, with 75 of its trucks involved in the operation, is one of four contractors charged with managing the entire WTC cleanup, an effort which ultimately will remove 1.2 million tons of debris.

Nor are many people aware of the efforts of **Tilcon New York, Inc.,** a division of **Oldcastle Materials**, which, working with fellow contractors **J. Fletcher Creamer & Son** and **Joseph M. Sanzari, Inc.,** organized a caravan of 150 pieces of equipment, trucks and quarry loaders brought to the site. Equipment operators from these firms worked 24-hour shifts, clearing streets of the more than 100 demolished automobiles, while other employees volunteered to work side-by-side

with police and firefighters in the bucket brigades.

Also on the scene immediately were employees of **Granite-Halmar**, the New York-based subsidiary of **California's Granite Construction**; of **Yonkers Contracting Company**; of **New Jersey's E. E. Cruz & Company** and **J. H. Reid General Contractor**; and of other companies too numerous to mention. New York City—and the nation— owe these hundreds of selfless people a debt of gratitude.

Meanwhile, engineering expertise has come from **Parsons Brinckerhoff** and the **Bechtel Corporation**, who immediately put 50 of their finest engineers to work assessing the structural damage to buildings in and around the WTC complex. Engineers from **Slattery Skanska** climbed through the collapsed and flooded subway tunnels and stations under and around the site, stabilizing and shoring up damaged tunnel zones in an effort to prevent further structural failure and flooding.

Another critical component of the cleanup operation came from **Jacobs Sverdrup Civil, Inc.**, which managed the construction of the new roadway used to haul debris from the site to the piers on Manhattan's west side, and from **Perini Corporation**, which constructed the road in record time.

At the same time, equipment manufacturers like **Caterpillar**, **Case-New Holland**, and **Ingersoll-Rand** supplied much of the huge and sophisticated equipment crucial to the dangerous dismantling and removal operations taking place at "ground zero."

And equally vital to the operation were the scores of other industry firms that helped out but who preferred no publicity for their actions or felt uncomfortable even being mentioned.

What happened at the World Trade Center cleanup is a good example of how the transportation construction industry steps up and does what needs to be done. But you don't need to look to New York—or wait for a time of national emergency—to find plenty of other examples. Because doing what has to be done—keeping America moving, creating and maintaining a transportation infrastructure without which we would live in perpetual emergency—is what our industry does.

And to celebrate that work is the purpose of the book you hold in your hands, a book being published as part of the American Road & Transportation Builders Association's (ARTBA's) 100th Anniversary of service to the U.S. transportation construction industry. As we consider the remarkable accomplishments of the transportation construction industry, it is instructive to recall that ARTBA's founder, Horatio S. Earle, may well have been the first to propose a national interstate highway network, articulating the concept in 1901. In February

1902, Earle formed the American Road Makers (ARM), known today as ARTBA, to push for federal legislation to create a "National Capital Connecting Highway System." His vision was accomplished with the 1956 enactment of the Interstate Highway Construction Program law—arguably one of the greatest association achievements in history.

The chapters that follow focus on a number of aspects of the achievements of the transportation construction industry—an industry that today generates $200 billion in economic activity and sustains more than 2.2 million American jobs.

These chapters show how the early years of the partnership between government and private industry led to the building of America's first improved roads. They show how, thanks to an expanding road network, Americans first began to discover and explore

the natural wonders of this great country. They highlight some of our individual road-builders—engineers and contractors—and trace the fascinating evolution of our road-building machinery. They examine the immense challenges facing the industry today and take a look at what the 21st century holds for our transportation infrastructure system.

Throughout the book one truth shines clear. Our transportation infrastructure is the thread that binds the fabric of America. The movement of people and goods is the world's core business. Transportation is the engine that drives the economy and that positively contributes to every aspect of our quality of life.

The title of this book asks, "Are we there yet?" Well, one thing is for certain—we are where we are because of our transportation infrastructure system. ∾

chapter one

INTRODUCTION

By Drew Lewis, U.S. Secretary of Transportation
under former President Ronald Reagan

 One quality American generations have shared from the very start is the drive to move, explore and expand the search for new opportunities. The narrow remains of the Santa Fe Trail near Las Vegas, New Mexico, 7,500 feet above sea level, recall Americans struggling westward for homes and liberty. No civilization can survive without means of reaching toward new goals. ∾ Our history is one of expansion— across or under rivers, over or through mountains—but no innovation or progress would have been possible without public and private participation and support— financially and politically. Financial institutions can raise funds, taxing authorities can demand funds, but the individual interests of federal, state and local governments are fundamental. Contractors and suppliers who move these plans into reality— many represented in the American Road & Transportation Builders Association— add unique strength. Growing the infrastructure of America today has indeed been a progress of consensus. It must continue being so in this new century. ∾ There can be no sitting on hands. Our transportation infrastructure has aged, and there is urgent need today for preventive maintenance, upgrading and new construction. ARTBA has the experience and talent to guide public and private expenditures for the right choices and the lowest costs. ∾ Nothing in this country has remained or will ever remain the same. Transportation 50 or 75 years into the 21st century may defy belief—to say nothing of today's technology. Just remember that the horse-drawn Conestoga wagon carrying pioneers to the West was the state-of-the-art transport of its day.

"Travel and transport by road are functions of civilized society."
—*John B. Rae,* The Road and the Car in American Life

The Public/Private Partnership That

Perhaps if our nation's capital had remained in Philadelphia, the federal government would have been slower to take up the cause of road building. After all, even before the end of the 18th century, you could travel from Pittsburgh to Philadelphia in only nine days. But in 1802, when Washington, D.C., had been the nation's capital for only one year, President Thomas Jefferson reported that the road from Fredericksburg, Virginia, to Washington was "the worst in the world," and seeds were planted for a partnership that still bears fruit today. ∾ Jefferson's vision was, of course, expansive, and building roads into the territories was fundamental to his political philosophy: Americans would follow these roads to find the inexpensive farm land upon which they would become the yeoman farmers of the Jeffersonian dream (Patton, 30). Congress responded to Jefferson's urging with the Enabling Act of 1802, the first legislation to commit federal aid to interstate road building. The act authorized an east-west route, which would become known as the National

Built America's Transportation Network

Road, running from the Potomac

River at Cumberland, Maryland, to the Ohio

River at Wheeling, West Virginia. ∿ But the debate over

Congress's authority to appropriate funds for "internal improvements"

was by no means settled. In 1822, President James Monroe vetoed an act that

would have preserved and repaired the National Road as it existed at that time. His

successor, John Quincy Adams, felt otherwise and sought to restore funding for this ven-

ture. Next came Andrew Jackson, who in 1830 vetoed a bill Congress had passed granting

federal dollars for the construction of the Maysville Turnpike in Kentucky. Jackson firmly be-

lieved that states should build their own roads (which Kentucky did), and yet, in Jackson's thinking,

the National Road was another matter altogether. This was an interstate undertaking, and therefore

could be federally supported. ∿ For the most part, however, it wasn't—especially after 1840, when the

emergence of the railroad ushered in the "dark ages" of road building. Although by 1850 the National Road

Evidence of the government's renewed interest in road building took the form of the Office of Road Inquiry (ORI) in 1893. The job of this entity, which would evolve into the Bureau of Public Roads (BPR), was to study methods of road making and to employ its findings in the construction of "object-lesson" roads. The importance of this work was magnified when, in the same year, Congress enacted Rural Free Delivery, which promised universal mail service…but only where gravel or macadam roads existed.

One hundred years ago, General Roy Stone, head of the Office of Road Inquiry, was using his position to revive the vision of a national system of highways—specifically "an Atlantic and a Pacific coast line, joined by a continental

The bicycling craze swept the country in the 1880s.

extended to Indianapolis, it was in such poor repair that even those who lived along it failed to take it seriously. Settlers took advantage of the cleared roadway to build their houses and, if the ground looked promising, to plant their gardens.

Ironically, given the beleaguered status of bicyclists on U.S. roads today, it was the advent of the bicycle that refocused the nation's attention on the condition of its roadways. The cycling craze soared in the late 1870s, and by 1880 cycling enthusiasts had formed the League of American Wheelmen (from whose seeds would sprout the American Road Makers [ARM] and, ultimately, today's American Road and Transportation Builders Association[ARTBA]). In a relatively few more years, the bicycle would give way to the automobile, and American roads would never be ignored again.

highway extending from Washington (D.C.) to San Francisco." At the same time, the American Road Makers was officially chartered, with the intention to carry forward the work of the "good roads movement" initiated by the League of American Wheelmen 20 years earlier.

Indeed, ARM's founder, Horatio S. Earle—who had also served as president of the League of American Wheelmen—had a vision every bit as large as General Roy Stone's. ARM's earliest goal, even at its founding in 1902, was for a "Capital Connecting Government Highway," which Earle foresaw as "the Eighth Wonder of the World." And Earle, a road commissioner from Michigan, infused his new organization with a sense of the potential inherent in a partnership between the road building industry and local, state, and federal government. As early as 1907, Earle had drafted ARM's National Reward Road Bill (which would have reimbursed states with federal dollars per mile of road construction) and was lobbying Congress for its adoption. Never a small thinker, Earle proposed for his Road Bill a $100 million appropriation from the Treasury—$10 million for each of the next ten years. (*Autobiography*, 112)

The membership fee [for ARM members] will be five dollars and the yearly dues five dollars. This is not a cheap affair, and cheap members will not be solicited."
—*Horatio Earle,*
THE AUTOBIOGRAPHY

A telegraph boy on his bicycle in Danville, Virginia, circa 1911.

In 1910, ARM incorporated as the American Road Builders Association (ARBA). Writing in his autobiography some 15 years later, Earle confidently claimed that the organization he had founded "has wielded a mighty influence in the land and, without doubt, has been the principal factor in winning the national battle for better roads. From a small membership, with comparatively little influence, it has developed into the most powerful organization of its kind in the land." (116) Indeed, from its inception in 1902, the association has played an integral role in the formulation and implementation of national transportation policy.

Rural postal delivery takes a lunch break in North Dakota.

In 1912 Congress passed the landmark Post Office Appropriations Act, which made available $500,000 for improving roads in several states in which rural free delivery of mail "is or may hereafter be established." This act not only spurred road construction on a national basis, but also paved the way for much more important legislation on the horizon. In 1916, President Woodrow Wilson signed into law the Federal Aid Road Act, considered the first real step by the federal government to establish an integrated, nationwide system of interstate highways.

ROUTES OF HISTORIC ROADS

LEGEND

PONY EXPRESS AND OVERLAND TRAIL
BOSTON POST ROAD
FORT DEARBORN TRAIL (CHICAGO TO DETROIT)
MAYSVILLE TURNPIKE (KENTUCKY)
EL CAMINO REAL
NATIONAL PIKE
ERIE CANAL TOW-PATH
THE OREGON TRAIL
THE CAMEL EXPRESS
FLYING MACHINE ROUTE (PHILADELPHIA TO JERSEY CITY)
WILDERNESS ROAD AND VALLEY TRAIL
ATLANTIC SEABOARD (PHILADELPHIA TO SAVANNAH)

This milestone legislation appropriated $75 million to be allocated to the states over the next five years for the construction of rural public roads—i.e., "any public road over which the Unites States mails now are or may hereafter be transported." In addition, the act had a number of interesting provisions:

★ FEDERAL CONTRIBUTIONS WERE NOT TO EXCEED 50 PERCENT OF THE TOTAL COST OF ANY PROJECT.

★ TO IMPROVE AS MANY MILES OF ROADS AS POSSIBLE, NO MORE THAN $10,000 IN FEDERAL AID WOULD BE SPENT PER MILE.

★ TO ALLAY FEARS OF UNDUE FEDERAL INTRUSION, THE STATES THEMSELVES WOULD BE SOLELY RESPONSIBLE FOR MAINTAINING ANY ROADS CONSTRUCTED UNDER THE PROGRAM.

★ IN ORDER TO QUALIFY FOR FEDERAL AID, THE STATES HAD TO HAVE AN OFFICIAL HIGHWAY DEPARTMENT STAFFED WITH TRAINED HIGHWAY ENGINEERS.

Workers spreading asphalt on the Boonsboro-Shepardstown Pike in Maryland.

Tough roadbuilding work in Nevada's Black Canyon, 1933.

Scenic highway built by the WPA atop Mt. Scott in Oklahoma.

Workers encamped on a Bureau of Public Roads project in Wyoming, 1920.

America's entry into World War I, along with a host of bureaucratic problems, drastically slowed the implementation of the Road Act. In fact, by March 1919 only 13 miles of federal-aid highway had been constructed (*The States and the Interstates*, 8), and by the end of the war less than $500,000 of the funds authorized by the law had actually been paid out to the states. Indeed, the program's apparent failure cast a cloud of doubt on the viability of such a state/federal partnership.

But as the end of hostilities in 1919 ushered in the boom times of the Roaring Twenties—to say nothing of the tremendous surge in automobile manufacturing in particular—road building wasn't an option; it was an urgent mandate of the public will. At this critical juncture, Thomas H. MacDonald left the Iowa Highway Commission to assume leadership of the Bureau of Public Roads—predecessor of the Federal Highway Administration—a post he would hold for the next 34 years. MacDonald saw the promise of the 1916 Road Act, along with its inherent problems. He had been a member of the American Association of State Highway Officials

(AASHO), and he was convinced that the national interest would be best served by a coalition that included AASHO, ARBA's membership, and federal policy makers. Those who criticized the 1916 law on the grounds that it lacked focus and direction had a point, but the solution was not to throw the baby out with the bath water.

Thanks largely to MacDonald's consensus building, when the Federal Aid Road Act was rewritten and re-signed in 1921, it included a crucial new provision: states were required to designate a specific system of roads, *not to exceed 7 percent of their total highway mileage*, upon which the federal monies would be spent. The selection of these roads gave focus to road building efforts, channeled federal dollars to their best use, and played a vitally important part in the early development of the interstate highway system. As a result, the Highway Act of '21 proved tremendously effective, and road construction soared along with the national prosperity of the 1920s.

The growing interest in and commitment to improving the nation's roads during the '20s is underscored by the surging popularity of ARBA's "Road Show" during the decade. This roadbuilders' exposition, which had taken place every year since 1909, came into its own now, as industry leaders felt new confidence in their partners at the federal level.

1957 Road Show held in Chicago.

The 1926 Road Show, held in Chicago's Coliseum, boasted a record 295 exhibitors, who showed off the latest in heavy and light equipment and new construction materials. The 15,000 attendees included highway engineers and officials, contractors, and manufacturer's representatives—the men who would oversee the improvement of roughly 100,000 miles of the Federal-aid road system between 1922 and 1929. (Kuennen, *Transportation Builder*, March 2001, 19)

ARBA's 1925 Road Show held in Chicago.

Federal-aid project on U.S. Highway 10 along the Jefferson River near Butte, Montana, 1935.

Uniform signage appearing on Ohio's U.S. 40 in 1925.

The crash of 1929 and the depression years that followed gave the federal government a unique incentive to become even more involved in the important work of building the nation's roads. Because President Franklin D. Roosevelt's National Recovery Act (NRA) conceived of highway work as a way to provide jobs for the vast numbers of unemployed, federal money was extended to road projects far beyond the provisions of the 1921 act. For example, federal aid was now used to extend state roads to and through cities, and the NRA provided for the improvement and landscaping of

Lake Shore Drive looking north, Chicago, Illinois, 1940.

existing roadside mileage—making roadways not only more attractive but much safer as well.

The Hayden-Cartwright Act of 1934 continued the road construction efforts of the NRA by appropriating $200 million specifically for the nation's highways and thereby reestablishing the temporarily lapsed federal-aid highway program. The act also contained a landmark provision, vigorously championed by ARBA, which foreshadowed legislative battles yet to come: federal largess would be withheld from any state that sought to divert the funds to non-highway-related purposes.

The federal government confirmed its commitment to a national transportation network when the Federal-Aid Road Act was rewritten in 1938. The new revision directed the Bureau of Public Roads to investigate the feasibility of build-

ing three multiple-lane highways across the country east to west and three more north to south. The bureau's study, issued in the report known as *Toll Roads and Free Roads*, drew up the following master plan for the nation's highways:

★ CONSTRUCTION OF A SYSTEM OF INTERREGIONAL HIGHWAYS WITH CONNECTIONS THROUGH AND AROUND CITIES.

★ MODERNIZATION OF THE FEDERAL AID HIGHWAY SYSTEM.

★ ELIMINATION OF HAZARDS AT RAILROAD GRADE CROSSINGS.

★ IMPROVEMENT OF SECONDARY AND FEEDER ROADS CONSISTENT WITH LAND-USE PROGRAMS.

★ CREATION OF A FEDERAL LAND AUTHORITY TO ACQUIRE, HOLD AND SELL LANDS FOR HIGHWAY RIGHT OF WAY.

The following year, in September 1939, Germany invaded Poland, and America's eventual entry into World War II following the attack on Pearl Harbor had the effect of restricting national road building to essential defense-related projects. But even before December 1941, President Roosevelt had appointed the National Interregional Highway Committee, and this committee's historic report, which was submitted to the 78th Congress as House Document No. 379, ultimately took shape as the Federal-Aid Highway Act of 1944. This legislation called for the selection of a national system of interstate highways, not exceeding 40,000 miles, connecting the important cities and industrial centers of the country and serving the national defense. It also appropriated a record amount of federal aid—$500 million a year for 1946-47-48, to be disbursed according to a special new formula: 45 percent of the total for primary highways, 30 percent for secondary, and 25 percent for urban connections (the first time federal funds had been earmarked for urban areas).

It's likely, however, that before framing the Highway Act of 1944, legislators had thoroughly digested the pamphlet *Sound Plan for Postwar Roads ...and Jobs*, published in 1943 by ARBA president Charles Upham. This document, which convincingly underscored the economic and social benefits of postwar highway building, was widely reviewed by lawmakers, government agencies, and transportation industry members. Such was its effect that it was reprinted five times, totaling 50,000 copies. (Kuennen, *Transportation Builder*, May 2001, 16)

Still, despite the energy and resources behind it, road building in the postwar boom years was hard-pressed to keep up with the incredible surge in the manufacture and sales of auto-mobiles. During the 10 years immediately following the war, approximately 50,000 miles of new roads were constructed. However, during that same period, enough new automobiles were constructed to stretch 200,000 miles, bumper to bumper. Something had to give, as the 84th Congress acknowledged in its passage of the Federal-Aid Highway Act of 1956.

After passage of the 1956 Interstate Highway Act, construction begins in St. Charles, Missouri.

Car ownership soared during the years immediately after World War II.

federal policymakers asked ARBA to investigate the inventory of construction workers available for such a project as well as the ability of the engineering profession to undertake the program. In response, ARBA produced five separate reports on the highway construction industry's readiness to respond to the call, documents which subsequently became key guides to implementing the program. What's more, to answer doubts as to the necessity of a network of limited access, high-speed highways, ARBA produced a film highlighting the new system's benefits, *We'll Take the High Road*, which was shown to civic groups around the country. The course was set upon the completion of what has come to be accepted as the greatest public works program in the history of the country.

ARBA's film, "We'll Take the High Road," highlighting the benefits of transportation investment, was shown around the country.

This far-reaching piece of legislation differed fundamentally from its predecessors in that it proposed to *complete* the interstate highway system—with the creation of the Highway Trust Fund-appropriated funds, accordingly—rather than simply continuing indefinitely to help states improve their highway systems. The seriousness of the federal commitment to this project was underscored by a radically revised funding ratio. While the Highway Act of 1954 had already increased the federal government's share of road construction costs to 60 percent (against 40 percent from the states), the new law authorized the federal government to pay 90 percent of the cost of interstate highway projects. Thus was born the Interstate Era, called by some the "Golden Years of Road Building."

ARBA's behind-the-scenes work in the creation of the 1956 legislation demonstrates the effectiveness of the public/private partnership that has continued to serve the nation's vital transportation interests. As the bill was taking shape,

THE INTERSTATE HIGHWAY SYSTEM PROJECT—HOW BIG IS IT?

★ Enough earth was moved for the Interstate System to cover Connecticut knee deep in dirt.

★ Enough concrete was poured to build a sidewalk from the earth to a point in space five times the distance to the moon.

★ To site the Interstates, highway authorities had to acquire land roughly equal to the total acreage of Delaware.

★ Enough drainage culvert was laid beneath the Interstates to handle all the needs of a city six times the size of Chicago.

Source: Tom Lewis, *Divided Highways*

Congestion comes to Houston's Southwest Freeway, 1972.

Oregon's I-5 showing signs of 1973's energy crisis.

As the Interstate system steadily progressed throughout the '60s, however, the national vision became clouded. Shrill voices began to decry road programs as antagonistic to nature, the environment, and the "quality of life." Old road men like legendary engineer Frank Turner, who, in his words, "were environmentalists from way back," were befuddled by the hue and cry, just as they were surely amazed when a citizens group in San Francisco stopped the completion of the Embarcadero Freeway in 1966, thereby rejecting $240 million in federal aid. Suddenly, the rules of the game were changing.

Blown in on these shifting winds were the National Environmental Policy Act of 1969 and 1970s Clean Air Act, among others; and, shortly thereafter, federal transportation policy was thrown into further turmoil by the Arab Oil Embargo of 1973. With energy-conscious consumers now burning less gasoline, fuel taxes levied per gallon were producing less revenue for the Highway Trust Fund—and this at a time

when more constituencies were clamoring for those dollars. In its 1978 reauthorization of the federal aid highway program, Congress for the first time addressed both highway and mass transit programs in the same piece of legislation—the Surface Transportation Assistance Act. ARBA's readiness for this change in focus was reflected by its name change of a year earlier to ARTBA—the American Road and Transportation Builders Association.

When the federal-aid highway program came due for reauthorization in 1982, so many cars were crowding so many roadways that President Reagan (no friend of tax increases) signaled his willingness to raise highway user fees. After a legislative battle that lasted all year, the new bill was finally passed and signed by the president in January 1983. It increased the gas tax by five cents per gallon and, for the first time, specifically set aside one penny of the tax to support mass transit programs. Enthusiastically supported by ARTBA, which had served as a mediator during the arduous process, the bill was a major victory for transportation interests.

But in the new age of competing "special interests," the public interest was no longer so easy to discern—or to serve. When the highway and mass transit programs came up for reauthorization in 1986, the legislation became so contentious—largely because of "demonstration projects" championed by individual members of Congress—that the bill remained unfinished at the end of the term, even though funding for surface transportation programs had expired on September 30 and construction had come to a virtual standstill. When Congress reconvened in 1987, ARTBA's "Highway Action Campaign" swung into high hear, sponsoring news conferences and industry rallies around the country and organizing a "Call Congress Day," along with a petition drive, in March as final congressional action loomed. President Reagan vetoed the bill, but his veto was narrowly overridden by Congress, and the frozen federal funds were finally released.

In 1989, with the initial construction of the Interstate system drawing to a close, and with a major overhaul of federal surface transportation programs looming in a time of stiff competition for limited financial resources, the transportation industry looked for a rallying cry. ARTBA responded with its "Building a Better America Through Transportation (BABATT) Campaign." This intense lobbying effort on behalf of better transportation for the country argued the necessity of a hike in federal fuel taxes to finance a significant increase in transportation spending.

Throughout the year Congress debated the legislation that would become the Intermodal Surface Transportation Efficiency Act (ISTEA). The House version of the legislation adopted ARTBA's fuel tax proposal—a five-cents-per-gallon increase which the House Public Works & Transportation Committee called a "Nickel for America." But the battle was far from over.

The question became, Who would get the nickel? The Congress was laboring under immense pressure to reduce the federal budget, and transportation advocates worried that fuel tax increases would go toward deficit reduction rather than to highway programs. When the House finally suggested a compromise—2.5 cents to the Highway Trust Fund and 2.5 cents toward the deficit—the transportation industry redoubled its efforts to ensure that the entire tax increase returned to the roadways.

CENTS

COALITION FOR AN EFFICIENT NATIONAL TRANSPORTATION SYSTEM

*It Makes Sense...
It Will Take 5 Cents*

The "Nickel for America" program ultimately became so mired in political squabbling that it was withdrawn by its House sponsors, but the idea behind it—the clear need for more investment in the nation's transportation system—was ultimately written into the bill signed by President Bush in December 1991. This piece of legislation, massive and complex, thoroughly revised federal highway law for the first time since 1916. Thanks at least in part to ARTBA and its allies, $155 billion was appropriated for the next six years, money that would cover additional revenues for the Highway Trust Fund extended past

Eastern span of the westbound lanes of the San Francisco-Oakland Bay Bridge, 1996. A new bridge is now under construction.

Aging infrastructure: Montpelier, Vermont's Pioneer Street Bridge will have to come down.

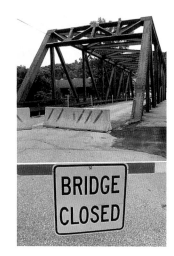

its scheduled expiration date, more research funding, preservation of safety programs, and major incentives for public/private ventures in transportation.

Several other provisions brought reassurance that the longstanding partnership between the federal government, the states, and the transportation industry was still in force: establishment and funding for a new, 155,000-mile "National Highway System;" an expanded bridge repair and replacement program; and a minimum 90 percent return to states of fuel taxes sent to Washington.

Looking ahead, ARTBA in 1995 invited national construction, general business, labor, tourism, energy, agriculture and modal groups to join it in the formation of the Alliance for Truth in Transportation Budgeting. The Alliance's mission: to push legislation to take the four federal transportation trust funds off-budget to eliminate revenue diversion and protect programs from future budget cuts. More than 100 organizations eventually participated. The following year, as the surface transportation program neared its reauthorization, ARTBA further focused its lobbying effort by initiating and co-chairing the Transportation Construction Coalition (TCC). This group provided the unprecedented vehicle to allow all facets of the transportation construction industry to speak with one voice in the pursuit of an industry-specific agenda in the upcoming legislation. Twenty-eight construction associations and labor unions joined the coalition.

The influence of these two groups was undoubtedly felt in 1998's Transportation Equity Act for the 21st century (TEA-21), which boosted surface transportation spending another 40 percent over five years. Among its programs are a $28.6 billion authorization for the National Highway System, $23.8 billion for interstate highway maintenance, $33.3 billion for the Surface

Transportation Program, and $20.4 billion for bridge reconstruction and repair. Moreover, the Alliance for Truth in Transportation Budgeting successfully petitioned TEA-21 to establish budgetary "firewalls" to guarantee that virtually all incoming federal gas tax revenues are invested for their intended purpose—improving the nation's surface transportation system.

The 21st century will bring its own urgent battles, and the country's transportation network will reshape itself to meet constantly changing needs. But one thing won't change: people will still demand to get where they have to go.

As we contemplate the problems and the possibilities inherent in our evolving system of transportation, it is instructive to bear in mind the remarkable achievement of the 20th century. The vision expressed by Horatio Earle's words of 75 years ago penetrates even to this day: "Good roads are indicative of a high state of civilization; they bespeak for the county, state, or nation, where they exist, an interest in education, religion, and the commonweal. They save a portion of the cost and facilitate transportation; they make centralized and better schools possible; they make it easy to get to church, library, club, grange, lodge, and far from least, to socialize the countryside, practically, with the help of the automobile, [and] turn the country into one big village." (145)

But, of course, the ongoing partnership between our government and our transportation industry has done a great deal more than build roads. It has, in fact, created a tremendously complex transportation infrastructure that quietly but efficiently meets the needs of the most advance civilization on earth. Highways, railways, airports, seaports, mass transit facilities, and, increasingly, intermodal connections among these systems—all these elements together make up America's remarkable transportation network, a network that provides our citizens with a freedom of mobility unimaginable a century ago. The partnership responsible for this achievement remains clear in its mission: to serve the public interest.

Heavy traffic and uncrowded bus lanes on Virginia's Shirley Highway.

Intermodal: Newark International Airport, the Jersey Turnpike, and a freight train loaded with containerized cargo at Port Newark.

chapter two

INTRODUCTION

By Thomas J. Donohue, President & CEO, U.S. Chamber of Commerce

The first century of America's transportation system has been an unqualified success. Our highways, bridges, railroads, airports, and waterways have generated unprecedented economic growth, spurred personal freedom and mobility, and efficiently channeled commerce to every corner of the world. This chapter vividly describes the positive and pervasive role the nation's transportation system has played in our daily lives and the remarkable standard of living we enjoy. ∾ But what about the second century of American transportation? Will it bring us as far and as fast as the first 100 years? What must we do now to ensure that our transportation infrastructure continues to be an engine of economic growth in the future? ∾ Today our network of roads, rails, runways, ports, and waterways is overburdened and in disrepair. Since 1970, vehicle miles traveled have increased more than 125 percent while road capacity has increased only by 6 percent. Road congestion costs the economy $72 billion annually and is a significant contributor to air pollution. ∾ In just the past five years, the number of passengers on U.S. airlines has increased 27 percent to more than 655 million travelers. Since the beginning of jet service, only two major hub airports have been built in the U.S. The other modes of transportation face similar problems. ∾ Our ability to create and maintain a first-rate transportation system will determine America's future global economic leadership. Without adequate infrastructure, there is no trade, no growth, and no competitiveness for America. ∾ Tell your congressmen, state representative, and opinion leaders in your community to invest in America's future by investing in infrastructure. Let's make the second 100 years of transportation even better than the first.

*"It was my deep-seated conviction that good roads were needed more,
as the solution of an economic problem, than anything else."*
—Horatio Earle

Economic Lifeline: How Transportation Has

From a distance, America in 1900 would look quite familiar to those who have just witnessed the arrival of the 21st century. ∾ In the presidential election that year, William McKinley, the Republican, was smeared by his opponents as the pawn of "big business." Meanwhile, William Jennings Bryan, the Democrat, ran as the champion of the working man, the people's candidate. (As in 2000, the Republicans prevailed.) ∾ Also in 1900, American troops were being dispatched around the globe, trying to quell uprisings in both the Philippines and China. The United Mine Workers went on strike, shut down the coal mines of northeastern Pennsylvania for six weeks, and, with the extra leverage of the upcoming election, ultimately won a 10 percent wage increase. Immigrants poured into the country in 1900, looking for a higher standard of living, a better life. Just as it does today, this influx had the paradoxical effect of threatening the comfort and security of native-born Americans while burnishing the nation's image as the land of of opportunity. ∾

Helped Create the American Marketplace

One hundred years ago, terrorism was

a frightening reality. King Humbert the First of Italy

was assassinated by anarchists in 1900, and attempts had been

made on the lives of the Prince of Wales and the Shah of Persia. Rumors

circulated that President McKinley himself was on the anarchists' "hit list."

And, like it is today, racism in 1900 was a familiar and divisive issue. Emerging from

Reconstruction, the South fell into the relentless grip of "Jim Crow," who would squeeze

from the "freed" black population all the rights their emancipation had promised them. ∿

Given the long view, then, the two turns-of-century have much in common. But as the camera

moves closer, the two images pull apart, showing worlds so divergent as to become scarcely recogniz-

able one to the other. What has changed, fundamentally, is our *way of life*. World events and national

issues notwithstanding, Americans in 1900 were a rural people, whose daily lives were still much as

they had been a hundred years earlier. ∿ Consider the routine of a turn-of-the-century Ozark farmwife,

Farmhouse in rural Custer County, Nebraska, 1889.

as recreated in the pages of Lebanon, Missouri-based *Bittersweet* magazine. This woman's first chore, upon awakening before daylight, was to bring in the wood to start the fire in her cookstove. While making breakfast—eggs from her own chickens, bacon or sausage from the smokehouse, biscuits from flour purchased in 25- or 50-pound sacks—she would have also carried in water from the well, which would be heating on the stove for coffee, and used later for dishwashing purposes.

Unless it happened to be the occasional Saturday that the family went to town for supplies, she would not leave the farm. She had plenty to do at home—cooking, cleaning, sewing, clothes-washing, gardening, and myriad other chores, all of which were made labor intensive by her utter lack of "conveniences"—no hot or cold running water into the kitchen, no refrigeration, no electricity. Whatever comforts she managed to introduce into this hard life—a glass of cider perhaps, or even apple wine—would have been literally "homemade."

More intimately, here is how Rose Wilder Lane, interviewed in 1940 as part of the Federal Writers' Project, remembered her childhood in the Missouri Ozarks at the dawn of the 20th century. After seven years of struggle, her family had finally abandoned the Dakota Territory during the panic of 1893:

Improved road winding through Maryland farm country, circa 1930.

seemed to have less and less allure. According to cultural historian James Flink, it was widely feared during the first decade of the 20th century that the migration of the rural population into cities would soon deplete the number of farmers to the point that a critical food shortage would result. "Rising prices for farm products disturbed city consumers," Flink notes, "...yet the financial rewards of farming were still not sufficient to keep talented and ambitious rural youth tied to a life of isolated drudgery." This grim picture was brightened by good roads and automobiles, which "promised to break down the isolation of rural life, lighten farm labor, and reduce significantly the cost of transporting farm products to market, thus raising the farmers' profits while lowering the food prices paid by city consumers." (40)

Indeed, the nation was poised on the brink of an economic revolution defined by the rapidly developing ability to *move goods*. Inside the cities, for example, many people felt that the trolley and the bicycle already offered themselves plenty of mobility, but their goods—once unloaded at the depot—were still moving through the streets on

Heavy traffic—by foot, wagon and trolley—clogs a cobblestone street in Philadelphia, 1897.

horse-drawn wagons or human-powered push-carts. Clay McShane cites an 1896 article in *Engineering News* that "complained of the high cost of intra-urban traffic and made the incredible claim that railroads could carry some classes of intercity freight for as low as twenty-five cents a ton/mile while urban freight cost up to twenty-five cents a pound/mile. Such high transfer costs figured to raise the price of consumer goods considerably." (120)

Outside the city, the tremendous improvement in rural roads initiated by Rural Free Delivery in 1896 meant that farmers and their wives would soon be pouring through the pages of the Sears Roebuck and Montgomery Ward catalogs. As roads continued to improve, the goods that these farmers ordered would actually be able to be delivered.

Good roads and the automobiles that demanded them moved people, too, of course, with

an economic impact that rippled ever outward. Those with the most immediate stake in the changes afoot were quick to tout the advantages inherent in the transportation revolution. In the early years of the new century, motoring periodicals were already claiming that doctors who drove could quadruple their house calls, and presumably their incomes, while sales agents could double the size of their territories. (McShane 121) In that light, it's instructive to look at the recollections of a North Carolina country doctor of the era, who, on the one hand, already romanticizes the days gone by, but, on the other, fervently wishes that things would get better quicker. He tells the story of falling asleep in the saddle of his white mule, Buncombs, after a long day of doctoring:

I GUESS THAT MULE HAD THE INSTINCT OF A HOMING PIGEON, FOR WHEN I FINALLY OPENED MY EYES HE WAS NUDGING MY GATE, ANXIOUS FOR THE WARMTH OF THE STABLE AND THE OATS HE KNEW AWAITED HIM THERE. YES, SIR, I REGRET THE PASSING OF THAT ANIMAL. . . .

NOW AS EVER THE COUNTRY DOCTOR MUST DEVOTE OVER HALF OF HIS TIME TO THE CARE OF THOSE UNABLE TO COME INTO TOWN FOR TREATMENT. AND IT SEEMS THAT NO MATTER HOW FAR WE HAVE COME FROM THE HORSE AND BUGGY DAYS OF A BYGONE ERA, OR HOW FAR WE HAVE PROGRESSED IN TERMS OF TRANSPORTATION AND GOOD ROADS, THERE STILL ARE TIMES WHEN THE COUNTRY DOCTOR MUST WALK MILES THROUGH ALL KINDS OF WEATHER OVER STORM-GUTTED PATHS, OR LEAVE HIS CAR STUCK FAST IN MUD HALF WAY TO HIS DESTINATION AND PROCEED THE REST OF THE WAY ON FOOT.

County road patrolmen maintaining New York roadway, 1914.

Wet weather rendered unimproved roads impassable.

LIFE IN THE 20ᵀᴴ CENTURY

1 in 7 homes had a bathtub
1 in 13 homes had a telephone
Brownie camera: $1
pound of sugar: 4 cents
dozen eggs: 14 cents
pound of butter: 24 cents
average weekly pay: $9.70

LIFE IN THE 21ˢᵀ CENTURY

VCR in 81% of homes
2.3 TVs per household
20% of U.S. connected to the Internet
pound of sugar: 43 cents
dozen eggs: $1.12
pound of butter: $2.35
average weekly pay: $435

If doctors and drummers could get out of the city to make calls in the countryside, by the same token farmers could get out of the country and into the city—and not merely for the purpose of bringing goods to market. "Get the farmer out of the mud" was the catchphrase, but as Phil Patton wryly observes, "the thing the farmer liked best about the car was that it would take him to town.... Thus the importance of the rural road. Justify it as they might as 'farm to market,' the improved rural road was also a route to town culture."(58) Broadly defined, of course, "town culture" included such

Iowa farm produce making its way to market, circa 1940.

Bad roads in Virginia bring good business to a farmer with a mule team.

What Depression? A freighter hauls a load of new cars through Michigan's Soo Canal, 1939.

Ford touring car, 1913.

fundamental amenities as education and medical care—with their even more profound and far-reaching effects on the economy.

By 1904, 154,000 of the nation's two million miles of roads had been "improved," and 10,000 had been paved. By 1916, when the first Federal Aid Road Bill was enacted by Congress, Henry Ford was producing 734,811 Model Ts a year. By the end of World War I, there was no looking back. (The dizzying pace of change during these years might be measured by a brief glance skyward: The Wright brothers went airborne out of Kitty Hawk, North Carolina, in 1903; Charles Lindbergh flew across the Atlantic in 1927.) In any case, as James Flink observes, "During the 1920s automobility became the backbone of a new consumer-goods-oriented society and economy that has persisted into the present." (141) The transformation of the nation's economy was well underway. The rapidly developing transportation infrastructure, by making more goods more available more quickly, would set the pace.

It is no exaggeration to say that, after the crash of 1929, the transportation construction industry kept the nation's economy afloat. Throughout the depression years, President Roosevelt rightly saw highway building not only as critical economic stimulus but also as a

Manual road maintenance in rural Idaho, 1935.

In the same year, an asphalt paver at work during Colorado's Columbia Basin Project.

Traffic Interchange System on the Cross Bronx Expressway.

solution to massive unemployment. Mark Foster, in *From Streetcar to Superhighway*, calculates that "Through mid-1939 the Works Progress Administration expended 38.3 percent of all its funds on roads, streets, and highways…[and that] of a grand total of 2,070,000 workers on WPA rolls at the end of 1939, over 900,000 were engaged in street or highway building projects." (168). According to FHWA's *America's Highways, 1776-1996*, congressional spending inspired by emergency relief programs pumped one billion dollars into highway construction between 1933 and 1938, money that financed more than 54,000 miles of improvement on the nation's roadways. (125)

Of course, on these roads people drove their cars, depression or no. Flink reports that motor vehicle miles of travel increased from 198 billion in 1929 to 206 billion in 1930 and 216 billion in 1931. At the depth of the depression, 1932-33, the total never dropped below 201 billion miles. Meanwhile special motor vehicle tax receipts progressively mounted throughout the depression years, from $849 million in 1929 to $1.69 billion in 1940. (160) The ultimate economic boon of all this travel—in the purchase of cars, car parts, and fuel, in the countless subsidiary industries that sprang up to serve the car and its driver—can scarcely be calculated. But one measure comes from economist Julius Weinberger, who estimated that vacation travel in the year 1935—85 percent of it by car—accounted for over half of the total estimated expenditures for all recreational purposes, $1,788 million out of $3,316 million." (Belasco 143)

With the radical shifting of economic gears at the end of World War II, the nation's roads paved the way into an era of peacetime prosperity. The economic impact of improved transportation infrastructure was certainly not lost on President Eisenhower when he signed the 1956 legislation to create the National System of Interstate and Defense Highways. As Phil Patton observes, "Eisenhower believed that highways would bring benefits for the whole economy. In speaking of the way better highways would provide 'greater convenience, greater happiness, and greater standards of living,' he echoed widespread public opinion." And after all, it was General Motors' Charles Wilson, Eisenhower's Defense Secretary, who said, "I've always believed that what was good for the country was good for General Motors, and vice versa."

It was a mindset that persisted up until that period of upheaval known as the sixties—a time when no cows remained sacred and no ideas were left unexamined. But if the last third of the century ushered in a era of unprecedented "consumer consciousness," it simultaneously ushered in a parallel rise in the number of consumer products to be conscious of—and an ever more sophisticated transportation infrastructure system to deliver them, culminating in a plethora of "overnight" and "same day" services standing by to consummate every e-commerce-driven desire. Will our transportation infrastructure network continue to "deliver the goods"?

OUR TRANSPORTATION INFRASTRUCTURE— HOW BIG IS IT?

Few of us fully grasp either the size and scope of our transportation infrastructure or the enormous economic impact of the transportation construction industry. Not many of us would guess, for example, that our road, rail, airport, transit, and waterway construction has a greater positive impact on the U.S. economy than the petroleum, motion picture and tobacco industries combined. A few facts and figures bring the picture into focus.

The U.S. transportation construction industry has built an increasingly intermodal transportation infrastructure that includes:
★ 3.9 million miles of American roads and highways
★ 5,400 American airports
★ 200,000 miles of U.S. freight and passenger railroad track
★ 5,800 miles of urban mass transit with more than 2,300 stations, and 360 ports.

★ The transportation construction industry generates $200 billion in annual economic activity and sustains more than 2.2 million American jobs.
★ Every $1 billion invested in transportation infrastructure generates more than $2 billion in economic activity and creates up to 42,000 American jobs.
★ Private sector transportation contractors employ 574,000 workers, who earn an average $15.33 an hour. An additional 600,000 people are employed in federal, state and local transportation departments. Firms that provide materials and services to transportation contractors provide another 960,000 American jobs.

(From: *An Economic Analysis of the U.S. Transportation Construction Industry*, by Dr. William R. Buechner.)

Bridge on coastal Highway 101 north of Los Angeles.

> "Transportation improvements have preceded every stage of industrial development in human history."
> —Walter L. Sutton and David Marks, HIGHWAYS AND THE NEW WAVE OF ECONOMIC GROWTH

Crown jewel of the Crescent is the Intermodal Container Transfer Facility (ICTF) at Seagirt Marine Terminal in the Port of Baltimore.

In "Highways and the New Wave of Economic Growth," FHWA's Walter L. Sutton and David Marks declare that "Transportation improvements have preceded every stage of industrial development in human history." At the beginning of the 21st century, the United States is entering upon what these authors describe as "a fifth wave of industrialization that is transforming the global market and changing traditional notions of development." Based on innovations in logistics and manufacturing, this new wave features product components manufactured in distant countries, then assembled into products near the point of their final consumption or use. Clearly, the key to this "fifth wave economy" is a fast and reliable transportation infrastructure network that minimizes the cost of production. The authors cite a recent report by a leading logistics company stating that nearly 80 percent of executives consider product delivery as important as product quality.

"Now, more than ever," write Sutton and Marks, "businesses require a seamless intermodal transportation system." And the transportation construction industry is responding. Consider, for example, the so-called "Chesapeake Crescent," stretching from Baltimore to Norfolk. "When goods arrive by air or sea," claim Sutton and Marks, "they can be shipped overnight on inter-

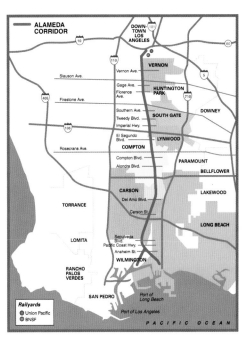

state highways or rail lines to more than 30 percent of the nation's population." Crown jewel of the Crescent is the Intermodal Container Transfer Facility (ICTF) at Seagirt Marine Terminal in the Port of Baltimore, which "raises intermodalism to an art form." The scant 100 feet from the bulkhead to the railhead insures that cargo arriving by ship speeds through the marine terminal to the rail yard, and on to the heartland of America.

On the other side of the continent Sutton and Marks find "one of the largest intermodal projects in American history": the Alameda Corridor, a 32-kilometer railway that will carry freight from the ports of Los Angeles and Long Beach to Southern California railheads. In addition to fostering California's fifth-wave business development, the Alameda Corridor is projected to support 700,000 new jobs in Southern California by 2020.

Such projects provide clear evidence of the transportation construction industry's determination not merely to keep pace with the economy, but—as it has always done—to create the space for economic growth. "In today's economy," Sutton and Marks conclude, "movement means improvement." True. But no more true today than when the first wheel turned its first revolution.

The Alameda Corridor plan area as shown on map.

Early piece of rail line snakes along Redondo Junction Bridge.

chapter three

INTRODUCTION

**By Ravi K. Saligram, Brand President, The Americas, Six Continents™ Hotels
(Owner and Franchisor of the Holiday Inn® Brand)**

For the past 50 years, Holiday Inn® hotels have led the evolution of U.S. road travel and the worldwide lodging industry overall. American entrepreneur Kemmons Wilson opened the first Holiday Inn hotel on August 1, 1952, in Memphis, Tennessee, after he returned from "the road trip that changed the world"—a family vacation during which he was discouraged over the lack of family and value-oriented lodging. ∾ Out of this frustration Wilson's vision emerged, and an American icon was born. ∾ Wilson expanded the brand, literally following the U.S. interstate highway system's bounding growth across the country. The familiar Holiday Inn sign soon came to represent comfort, quality, consistency, and value to millions of travelers, just as it does today. ∾ Now, with 1,600 hotels worldwide, Holiday Inn Hotels and Resorts is the most widely recognized lodging brand in the world. Every night, more guests stay at a Holiday Inn hotel then any other hotel brand in the world. Ninety percent of American travelers have stayed at a Holiday Inn hotel, more than any other hotel brand. The brand continues to offer the same dependability, friendly service, and modern, attractive facilities at an excellent value to today's travelers as it did in 1952. As the owner of the Holiday Inn brand, Six Continents Hotels, Inc., carries the brand's legacy as a fixture in American culture. Six Continents Hotels continues to honor the Holiday Inn brand's long-standing commitment to hosting travelers and is primed to take the brand into the next 50 years and beyond.

Almost from its beginning, the automobile was seen as a pleasure machine,
meant to increase social opportunities, to provide adventure and access to nature and, soon,
to a whole range of amusements created exclusively by the highway culture."
—*Phil Patton,*
Open Road

America the Beautiful: Opening Access to

In America in 1900, vacations were largely an indulgence of the wealthy, who by the turn of the century had established several elegant resorts, most of them on the eastern shore. ⌒ Martha's Vineyard, for example, was already a popular destination in the early 19th century, and by the latter part of the century 12,000 of the well-to-do vacationed there annually. In Newport, Rhode Island, the super-rich built some of the nation's most fabulous chateaux, where they idled away their summers in royal splendor. In the late 1800s, Coney Island was a fashionable resort, with grand hotels and manicured horse racing tracks—though the exclusivity of this get-away declined precipitously when New York City ran a subway line to the island in 1920. Inland were Saratoga Springs and the Catskills, among others, where the absence of the sea was compensated for by fresh air, mountain scenery, and healing mineral waters. ⌒ The railroad, of course, provided transportation to these destinations—as well as to more recently popular ones out west: Denver, Santa Fe, Los Angeles, San Francisco. In fact, as a way of increasing their

Our Historic Sites and Scenic Wonders

own trade, the railroads tirelessly

promoted whatever attractions lay along, or at the ends

of, their rail lines. The Maine Central Railroad, for example, touted

Maine as "the nation's playground," "a sportsman's paradise," and "the land

of the world's health." Of course, after a lengthy, bothersome journey with sched-

uling and other logistics dictated by the railroad, it made sense for the vacationer, once ar-

rived, to stay as long as possible. Day trips, overnighters, even weekenders—railroad

transportation made these for the most part unthinkable. 〜 So whatever the vacationer's desti-

nation in 1900, he needed money for transportation and accommodations, and he needed leisure time.

In what was still very much a rural, labor-driven economy, very few people had either. Fewer still had both.

But change was on the horizon. 〜 Two forces converged as the new century began. First was the "wilder-

ness craze" sweeping the country. The writings of John Muir were suddenly in fashion, and the Audubon Society

and Sierra Club were recently formed. Hiking and mountaineering became the enthusiasms of the moment,

STEAMBOAT SPRINGS
COLORADO
MOFFAT ROAD

Auto tourists heading through a rock tunnel along Wyoming's Wind River, 1924.

An AAA Good Roads official passing through Montana on a transcontinental trip, 1912.

Four women and a Packard: touring New Mexico's Enchanted Mesa in the 1920s.

By 1905, millionaire auto enthusiast Charles Glidden had already decided to present a trophy to the American Automobile Association (AAA), on which would be engraved the name of the owner-driver of the car making the best record on a carefully organized long-distance tour. The "Glidden Tour" would be an annual affair, supervised by the AAA, which would draw up the rules for the contest and retain ownership of the trophy.

A hugely popular event in its own right, the Glidden competition served a much more important purpose. According to the FHWA's *America's Highways*, "The Glidden Tours showed conclusively that motor cars were mechanically capable of traveling long distances if the roads were reasonably good, and they fueled a rising demand from motorists for motorable long-distance roads, even a coast-to-coast highway." (61-62) At the same time, by demonstrating the reliability of the moderately priced light car, such endurance runs opened the American mind to the possibility of the cross-country automobile vacation, thus inaugurating long-distance automobile touring by the average automobilist.

How rapidly this enthusiasm took hold is suggested by the fact that the 78,000 automobiles on America's roads in 1905 had by 1915 mushroomed to 2.33 million. And many of these new

as city people, already nostalgic for the lost frontier, fled to the great outdoors—especially to the five national parks already in existence.

Second, of course, was the advent of motor touring on the nation's slowly expanding network of improved roads. From the first years of the century up until the beginning of World War I, the national imagination seems to have been chiefly fired by stories of long-distance races and cross-country treks in the fabulous new vehicle—the automobile.

"Women drivers of motor vehicles should be given special consideration—and watching."
—from "Things Worth Knowing,"
1926 edition of RAND-MCNALLY ROAD ATLAS

motorists headed to the fabulous scenic wonders of the West. Clay McShane has noted that the American traveler's own innate desire to "go west" was often abetted by the media's fascination with the early automobile: "Many [newspaper photos of automobiles] displayed a conquest of nature motif. Cars perched at the edge of the Grand Canyon, traveled through a tunnel bored in a Sequoia tree in Yosemite National Park, or climbed Pike's Peak. The *Times* showed many photos like these and praised the motor car for its forays into deserts in the American west." (131) And indeed, travel historians generally agree about the huge imaginative appeal of early auto touring, wherein these intrepid motorists saw themselves as spiritually akin to the first pioneers rolling over the great plains.

One of the Loops, Highway to the Oregon Caves.

According to *America's Highways*, "The work in federally owned areas built up so rapidly that, by 1916, the OPR was maintaining 160 miles of road, constructing 170 miles and making surveys and plans for yet another 477 miles—a total program that was spread over 12 states and Alaska, and which exceeded the programs of a number of State highway departments." (75-76)

Their work was timely. Americans were eager to hit the road. Cars were becoming plentiful and affordable (Henry Ford's assembly line began cranking out the famous Model T in 1913), and new highways were opening up the country. In that same year, work began on the Lincoln Highway, brainchild of Carl Fisher and Henry Joy, and the spirit of the times was captured in the catchphrase popularized in their Lincoln Highway Association brochures: "See America first."

In the meantime, Congress in 1912 had required that 10 percent of the revenues from the national forests should be spent to construct roads and trails within these forests. Soon enough, as the forest road fund accumulated, the Forest Service was seeking help from OPR director Logan Page, asking his office to inspect and improve existing roads and to suggest where others should be built. The Secretary of the Interior, meanwhile, also asked for help with the planning of roads in the national parks. Page began in Yosemite, placing an engineer and a field survey party there during the summer of 1914, and promised assistance in the other four parks as soon as he could find the engineers.

Fisher, especially, understood that highways equated to tourism, and by 1915 his next project—the Dixie Highway—proclaimed itself "your favorite route to and from Florida or the Great Smokies." As Carl Patton observes, "Marked by red and white blazes on telephone poles, the Dixie Highway ran from Bay City, Michigan, through Chicago and Cincinnati, into Kentucky and Tennessee and around the peninsula of Florida in a great loop. It helped sustain the Florida land boom of the twenties . . . and it boosted local sights along the way." (44-45)

Forest Highway built by Bureau of Public Roads.

The Dixie Highway running through Davenport, Florida, 1923.

CARL FISHER: HIGHWAY DREAMER

DESPITE THE SEVERE astigmatism that forced him to quit school when he was twelve, CARL FISHER saw very clearly that the future lay along AMERICA'S HIGHWAYS.

IN 1891 AT AGE 17, HE OPENED A BICYCLE shop with his two brothers—just in time to take advantage of the national bicycle craze, which, in turn, focused interest upon the condition of the nation's roads. Then, as the bicycle began to yield its place to the next transportation phenomenon, FISHER cannily converted his bicycle shop into an automobile repair and sales facility. From there he moved to automotive innovation, joining patent-holder FRED AVERY to found the PREST-O-LITE company, which manufactured compressed gas headlights. (UNION CARBIDE would buy out the company in 1911 for the then-staggering sum of $9 million.)

A trip to FRANCE in 1905 convinced him that AMERICA needed a well-made testing ground, where U.S. cars could be raced, refined, and elevated to the level of the superior EUROPEAN models. He and three partners put up $250,000 to form the INDIANAPOLIS MOTOR SPEEDWAY COMPANY, which soon would become—and forever remain—synonymous with automobile racing.

A well-paved track was one thing. Out in the real world, FISHER realized, the automobile's potential was severely hampered by the nation's bad roads. In 1912, at a dinner party for automobile makers held in INDIANAPOLIS, FISHER unveiled his grand design: a "COAST TO COAST ROCK HIGHWAY"

stretching from NEW YORK TO CALIFORNIA. HENRY JOY, president of PACKARD, contributed $150,000 to the cause, along with a suggestion for the road's name: the LINCOLN HIGHWAY. Thus was born the LINCOLN HIGHWAY ASSOCIATION, with JOY as president and FISHER as vice president, and FISHER immediately set about to map the straightest feasible route from TIMES SQUARE to SAN FRANCISCO'S LINCOLN PARK. It would take another 12 years to complete the massive 3,389-mile project.

LONG before then, FISHER had turned his restless energy to yet another huge undertaking, one which, more than any of the others, demonstrated his understanding of the synergy between transportation and tourism. Ultimately, he would develop a dismal swamp at the southern tip of FLORIDA into a resort called MIAMI BEACH, but in order to do so, he would build yet another interstate highway. In 1914 he proposed to INDIANA GOVERNOR SAMUEL RALSTON the creation of a highway running from CHICAGO to MIAMI. RALSTON received eager support from the other governors along the route, and the "DIXIE HIGHWAY" was a reality by the end of 1916.

FISHER went on to develop MIAMI BEACH and was involved in a new development at MONTAUK on LONG ISLAND when the market crashed in 1929. He lost everything he had and died ten years later, virtually penniless, in MIAMI BEACH.

But along ROUTE 30 IN PENNSYLVANIA, or along ROUTE 25 in NORTH CAROLINA, on highways all across the country, travelers today still share in FISHER'S legacy—a legacy as long as the AMERICAN road.

By the end of World War I, America's passion for motor touring—and for "seeing America first"—was firmly established. Historian Warren Belasco reports that, in 1919, 40,000 people in 10,000 cars, representing 46 states, visited Yellowstone National Park—all part of the growing national obsession called motor camping.

The earliest campers thought of themselves as gypsies of the highway. Freedom from the strictures of urban life was their only thought. They packed up their autos and headed into the countryside, and at the end of the day's drive they pitched their tents wherever they happened to be—perhaps by a clear-running stream, often in a farmer's conveniently cleared field.

This sort of pure vagabondage carried its concomitant risks and worries, though, and a rapid roadside evolution was soon underway. The progress from unfettered "camping out," to the "free campground" provided by villages enticing tourist trade, to privately operated "tourist cabins" offering an increasing number of amenities, was steady and predictable, as motorists concluded that freedom and comfort were not mutually exclusive.

Wyoming Forest on the Wind River Highway, 1921.

Auto campers in California's Death Valley, 1949.

In any case, the numbers suggest a remarkable phenomenon. It is estimated that by 1921 nine million Americans were motor camping annually. By 1926, writes James Flink, "some 5,362 'motor camps' dotted the American countryside, and an avalanche of tourists who never before had traveled more than a few miles from home began to descend on distant national parks, forests, and points of historic interest." (156)

As suggested above, those who lived along the highway were quick to realize that tourism meant business, and seeing to—or expanding upon—the needs of these motorists soon became a growth industry. And if the travelers enjoyed picturing themselves as latter-day pioneers, or even

native Americans, roadside merchants were perfectly willing to enhance the illusion. Thus the appearance of wayside "tavernes" offering "heartye fare" and "olde-fashioned" hospitality, and tourist cabins in the guise of Indian wigwams.

And thus the quick transformation of the interior of the tourist cabin, which at first offered nothing but shelter and a reprieve from the chore of setting up and taking down a tent each day. But by the mid-'20s more elaborate units were already proliferating, which, for a dollar a night, might offer a bed with a straw-stuffed mattress, a bench and table, a water pitcher and bowl. "Despite their Spartan furnishings," writes Belasco, "these cabins often filled up before dark. This was a real breakthrough, marking the end of autocamping and the beginning of the motel industry." (131)

"Wigwam" tourist cabins offered the illusion of a vanished America.

Tourist cabins along U.S. 11 in Virginia, circa 1935.

The word "motel" was coined in 1925 in San Luis Obispo, California, by James Vail, whose Mo-tel, as he spelled it, is still in existence today.

As the automotive age moved into high gear, roadside entrepreneurship kept pace, often taking odd and novel forms. The wayside's potential as an advertising medium was quickly grasped, and billboards popped up with perhaps annoying predictability. But one roadside campaign, dating from the mid-'20s, left a lasting impression on the nation's consciousness. As cultural historian Christopher Finch tells the story, a successful Minneapolis insurance man named Clinton Odell was ordered by his doctor to find a less stressful line of business. Along with a chemist acquaintance, he made the curious choice of developing a brushless shaving cream, which he named for the country of origin of some of its ingredients. When marketing the product door to door proved every bit as taxing as the insurance business, Odell's oldest son, Allan, had the idea of installing signs alongside highways leading into the city: not conventional billboards but sequences of half a dozen small signs that together would make up a jingle advertising their product:

"DOES YOUR HUSBAND /
MISBEHAVE /
GRUNT AND GRUMBLE /
RANT AND RAVE? /
SHOOT THE BRUTE SOME /
BURMA-SHAVE."

"Burma-Shave" signs could still be found along the roadway in the 1950s.

Sales of the product soared, more firmly entrenching commercial signage on the American landscape. (90-91)

The lean years of the Great Depression did little to dampen either the American appetite for motor touring or the government's commitment to road building.

Indeed, during these years the federal government seized upon scenic road construction as a way to ease unemployment while continuing to whet the national thirst to "see America." The pioneer federal project, according to Phil Patton, was the Mount Vernon Memorial Parkway, begun in the late '20s. "In the Blue Ridge Parkway and Skyline Drive, in the Norris Freeway built by the TVA, the parkway functioned as a way to show off great landscape. Along the Natchez Trace Parkway, the landscape was tied to the history of the area: old portions of the original Trace, along with cabins and an inn, were preserved within its right of way." (71)

Nor could the depression cool the entrepreneurial fever that spread along the nation's highways. It was in the early '30s, in fact, that

series of Federal-Aid Highway Acts culminating in the Interstate Highway Act of 1956, ushered in an era of road building—and motor touring—unlike any preceding it. Soldiers came home, bought homes and cars, raised families, and, like never before, took them on vacations. As Patton observes, "People visited parts of the country they would never have seen before World War II—they wanted to see the country they had left to defend." But they weren't interested in auto-camping, or any kind of camping. The war had given them plenty of that. "They wanted not only unfamiliar sights to see," writes Patton, "but familiar lodging and eating establishments to resort to—the comforts they had given up to protect." (188) Fortunately, the roadway had prepared for them. After two decades of evolution, the auto camp had completed its transformation into the motor hotel.

What's more, thanks to the vision of men like Kemmons Wilson, postwar vacationers would soon be able to find familiar lodging from one end of the country to the other. Wilson, a Memphis builder who had had his own difficulties in finding satisfactory accommodations while traveling with his family, had a new idea for the old "tourist court": rooms large enough for two double beds, modern toilet facilities, a clean restaurant, Coke machines and free ice, a swimming pool and hotel-style services. He built his prototype in the early '50s, and by the time the Interstate building program was underway, his Holiday Inn Hotels had become the nation's first widely established motel chain. (Finch, 236-237)

This Alabama billboard dates from the late '30s.

The Holiday Inn "Great Sign" began to proliferate in the late '50s.

Howard Johnson, a soda fountain owner in Quincy, Massachusetts, persuaded a fellow restaurateur in Cape Cod to rename his establishment Howard Johnson's and purchase his supplies from the store in Quincy. The arrangement proved profitable. By 1935, 25 Howard Johnson's dotted the highways of Massachusetts, and by the end of the decade a hundred outlets covered the East from Maine to Florida—all sporting the unmistakable orange roof. Each store offered the same menu, the same clean restrooms, all the comforts of uniformity. (Finch, 163) The formula was successful, to say the least, but World War II would come and go before the franchising of America began in earnest.

As noted in an earlier chapter, the end of the second world war, in conjunction with the

Electric sign outside Howard Johnson's restaurant promoting HoJo's ice cream, etc. topped with trademark image of chef giving boy a plate of pancakes.

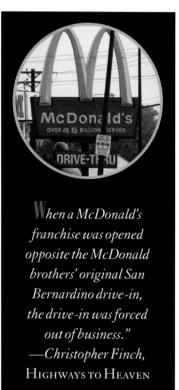

When a McDonald's franchise was opened opposite the McDonald brothers' original San Bernardino drive-in, the drive-in was forced out of business."
—Christopher Finch, HIGHWAYS TO HEAVEN

Given the amenities provided by the roadway entrepreneurs, and given the implications of the Interstate Highway Act of 1956, it is no wonder that the National Park Service and the Bureau of Public Roads (BPR) chose that same year to reaffirm their collaboration with the inauguration of MISSION 66. The ten-year program, wrote Conrad Wirth in *American Road Builder*, "aims to complete the physical developments and staffing requirements necessary to handle some 80 million visitors who are expected annually in the National Park System by 1966."

Others quickly followed suit. Historian John Rae reports that by 1960 Travelodge operated 110 motels; Howard Johnson's, having expanded into the motel business, had 89; and Holiday Inn had 160, with 15,000 rooms, putting it next to the Hilton and Sheraton hotel chains in size. (105-106)

Salesman Ray Kroc, meanwhile, had already asked himself why a burger joint in San Bernardino would need to order eight of his Multimixer milkshake machines—each one of which made five milkshakes at a time. The answer, he discovered, was the huge crowd of patrons who lined up for the restaurant's dependably good food, low prices, quick service, and attractively clean premises. After working out a deal with the McDonald brothers, Kroc began to sell McDonald's franchises in 1955, and within two years one hundred were in operation. As Christopher Finch observes, "The Golden Arches, [Holiday Inn] 'Great Sign' and Kentucky Fried Chicken's revolving tub became ubiquitous at exactly the moment in time that the new Interstate exit ramp strips were proliferating." (241)

With the program's target date coinciding with the 50th Anniversary of establishment of the National Park Service (and of its collaboration with the BPR), MISSION 66 also took the opportunity to celebrate some of the spectacular achievements of this long and fruitful partnership: the Going-to-the-Sun Road in Glacier National Park, the Trail Ridge Road in Rocky Mountain National Park, the Rim Drive in Crater Lake National Park and the Blue Ridge and Natchez Trace Parkways, among others.

ECONOMIC RIPPLES

Transportation construction is rightfully described as an "economic multiplier." Our transportation infrastructure has not only made it possible to get from one place to another; it has, in the process, given birth to the enormous travel and tourism industry. The economic impact of the availability of "bed and board" while we travel the nation's highways is suggested by the following statistics:

THE LODGING INDUSTRY
★ includes 53,500 properties, with 4.1 million rooms; *and*
★ generates $108 billion in annual sales.

THE FAST FOOD INDUSTRY
★ employs more than 2.5 million people; *and*
★ generates $107 billion in annual sales.

THE CONVENIENCE STORE INDUSTRY
★ operates 119,751 units; *and*
★ generates $269 billion in annual sales.

Or, looking at the big picture—

THE TOURISM INDUSTRY
is the nation's third-largest retail industry, generating $561 billion in sales annually (or $1.5 billion *a day*);
★ creates more than 7.8 million jobs, which pay out $171 billion in wages and salaries; *and*
★ pays $99.5 billion in federal, state and local taxes.

Clearly, the transportation construction industry "paves the way" for U.S. economic expansion.

This Utah highway winds through a natural tunnel in Bryce Canyon National Park.

Auto tourists pull over to admire buffalo in Missouri, circa 1938.

Some trends continue. Road building, motor touring, franchising, enjoying our nation's scenic wonders: these interrelated pursuits have become ingrained patterns of the American experience. The last link of the Interstate system was opened in 1991. As we enter the 21st century, 250 million people are visiting our national parks each year.

The doomsayers cry, "Enough!" How can the Park Service fulfill its stated mission—"to provide for the enjoyment of [the parks] in such manner and by such means as will leave them unimpaired for the enjoyment of future generations"—with 250 million tourists driving through them every year?

Is the solution to ban the vehicles and close the roads? Historian John Rae, for one, doesn't believe so. Noting that 95 percent of the visitors to national parks like Glacier, Great Smoky, Grand Canyon, and Yellowstone arrive by car and that, without such private transportation, few Americans could enjoy the scenic grandeur of such places, Rae writes that "the highway and the car have opened to multitudes of people new opportunities to travel for recreation and pleasure, and the volume of such travel will expand." (141)

Rae acknowledges that the preservation of our parks has become a matter of acute concern, but he argues that to blame the automobile, or to eliminate automobile access to such areas, is misguided. "The remedy," he writes, "is not to lock up recreational resources so that they can be enjoyed only by a minuscule (and largely self-selected) minority, but for public authorities at all levels to plan with foresight and intelligence for the preservation of these resources so that they will be of maximum benefit to our whole society." (142-143)

With such level-headed counsel in mind, it is reassuring to look at what's going on right now in Grand Teton National Park. Writing in a recent issue of *Public Roads*, Gary Hunter reports that 3 million visitors come to Grand Teton National Park every year— "to experience its vast array of wildlife, its outstanding beauty, its spirit of wildness. They come in sedans and station wagons and vans, in pickup campers and travel trailers and motor homes and tour buses. These vehicles need roads, and they need parking lots."

To meet these needs, the Federal Highway Administration (FHWA [formerly the BPR]) is designing and reconstructing the park's roads and supplementary facilities in such a way as to provide adequately for visitors "while preserving the park's wild—but delicate—beauty." With special consideration for maintaining the park's sense of wildness, the FHWA has set about the reconstruction and partial relocation of 8.7 km of highway, the overlay of 2.7 km of existing road, the construction of six parking areas, the obliteration and return to nature of 8.8 km of old road, and the rehabilitation into a natural area of a large borrow pit. The project's modest cost— $3,032,000—is also a wonderful harbinger. Such work can be undertaken in other national parks without great fiscal strain.

America's motorists are fortunate that her highways provide access to some of the world's most spectacular and breathtaking natural wonders. But with Yellowstone or Mount Rainier as our destination, let's not forget to enjoy the scenery along the way. However we may feel about the fast food and motel strips that lay at the foot of exit ramps, Christopher Finch reminds us that "thousands of miles of Interstate highway unwind gently across prairies, deserts and mountains...actually enhancing the motorist's ability to enjoy America." (228)

Thanks to the imaginative design and unsurpassed engineering of our nation's road builders, millions of motorists every day have the privilege and pleasure of seeing America first.

Into the sunset in the mountains of Colorado.

chapter four

INTRODUCTION

By General Barry R. McCaffrey, USA (Retired)

The American people need to understand and value the enormous contribution to our national military security as well as our safety from domestic natural disaster which we gain from our huge investment in our national network of highways, bridges, airports, railways, waterways, and ports. We cannot afford to take this complex and sophisticated capability for granted. This transportation infrastructure plays a critical role in our military preparedness and mobilization ability as well as our domestic emergency response to natural disasters. ∼ Thanks to America's unsurpassed transportation network, the U.S. Armed Forces has been able to deploy masses of troops and material to meet our global responsibilities during World War II, Korea, Vietnam, Desert Storm, and now the War on Terrorism. Similarly, in times of national emergency when American lives and property are threatened—whether the San Francisco earthquake of 1989, or Hurricane Andrew, or the tragedy of September 11 in New York and Washington—it is our transportation infrastructure that makes possible the great work of the Federal Emergency Management Agency (FEMA), the U.S. Armed Forces, the American Red Cross, and our brave Fire and Police personnel. ∼ We have created a national treasure through the creativity and hard work of America's road and transportation builders. In the coming 100 years we need to continue to invest our resources, leadership, and energy to maintain and adapt this system. Our safety and security depend on our continued dedication to this transportation responsibility.

General McCaffrey was former Drug Czar under President Bill Clinton and Commander of the 24th Mechanized Infantry Division during Operation Desert Storm in the Persian Gulf

"At the height of the Roman Empire, a map of hammered gold hung in the palace of Caesar, embossed with the lines of the Roman road system. . . . This map expressed a system that belonged to Caesar, built by his legions as the military and communication tendons holding his empire together."
—Phil Patton, Open Road

Ready in Crisis: Transportation's Role in

It's probable that no agency of public service is so taken for granted as our nation's transportation infrastructure. As a result of the increasingly thankless work of our transportation construction industry, millions of commuters, vacationers, motorists, and air travelers have come to expect quick and convenient transportation as if it were their birthright. And many, in this age of conflicted values, execrate our roadways even as they speed down them to their next urgent appointment. ∾ But throughout this century of revolutionary change, unprecedented growth, and constant innovation, the transportation construction industry's mission has remained constant: to serve the public interest. The industry's remarkable success, sometimes obscured by alternative agendas, becomes crystal clear in times of national urgency. This chapter, by outlining a series of occurrences – natural disasters and military operations—over the course of the century, dramatically highlights the impulse that has shaped the building of the transportation infrastructure we know today: *to be there when you need it.* ∾

Military Operations and Natural Disasters

At the turn of the century, Galveston, Texas, was a thriving coastal city. Its deep-water harbor made it the nation's busiest cotton port, and, after New York, it was the most popular entryway for immigrants from Europe. Its wide sandy beaches and mild Gulf waters lured not only locals down to the shore but railway tourists from as far away as Kansas City. ∾ It was still the summer season on the morning of September 8, 1900, and the abnormally tall waves were, if anything, an added attraction to the multitude of beachgoers who frolicked under increasingly cloudy skies. However, one man on the island, Isaac Cline, local representative of the National Weather Service, was disturbed by the conditions, and he promptly reported the "unusually heavy swells from the southeast" to the Washington headquarters. ∾ As the dark clouds, dense rain, and high wind began to besiege the city, the only warning of the nightmare to come came from Cline himself, who in the early afternoon harnessed his horse to a two-wheel cart and rode up and down

the beach, shouting, "Seek higher ground. Move into the heart of the city" . . . and from his office, where the telephone rang until the last line went dead.

It took Cline two weeks to file his official report with the National Weather Service. Here is a brief excerpt:

I REACHED HOME AND FOUND THE WATER WAIST DEEP AROUND MY RESIDENCE. . . . ABOUT 6:30 P.M. MR. J. L. CLINE [CLINE'S BROTHER] . . . HAD ADVISED EVERYONE HE COULD SEE TO GO TO THE CENTER OF THE CITY AND HE THOUGHT WE HAD BETTER MAKE AN ATTEMPT IN THAT DIRECTION. AT THIS TIME, HOWEVER, THE ROOFS OF THE HOUSES AND TIMBERS WERE FLYING THROUGH THE STREETS AS THOUGH THEY WERE PAPER AND IT APPEARED SUICIDAL TO ATTEMPT A JOURNEY THROUGH THE FLYING TIMBERS. MANY PEOPLE WERE KILLED BY FLYING TIMBERS ABOUT THIS TIME WHILE ENDEAVORING TO ESCAPE TO TOWN. . . .

[AT] 8:30 P.M. MY RESIDENCE WENT DOWN WITH ABOUT 50 PERSONS WHO HAD SOUGHT IT OUT FOR SAFETY, AND ALL BUT 18 WERE HURLED INTO ETERNITY. AMONG THE LOST WAS MY WIFE WHO NEVER ROSE ABOVE THE WATER AFTER THE WRECK OF THE BUILDING. . . .

More than 6,000 were killed and 10,000 left homeless from Galveston's "Great Storm."

The receding waters left a scene of total devastation.

SUNDAY, SEPTEMBER 9TH, 1900, RE-VEALED ONE OF THE MOST HORRIBLE SIGHTS THAT EVER A CIVILIZED PEOPLE LOOKED UPON. ABOUT 3,000 HOMES, NEARLY HALF THE RESI-DENCE PORTION OF GALVESTON, HAD BEEN COMPLETELY SWEPT OUT OF EXISTENCE AND PROBABLY MORE THAN 6,000 PERSONS HAD PASSED FROM LIFE TO DEATH DURING THAT DREADFUL NIGHT.

I. M. CLINE
GALVESTON, TEXAS
SEPTEMBER 23, 1900

The condition of thousands of those who have been spared is far more pitiable than that of the dead. Their resources have been swept away by wind and tide, and they are desolate in the midst of desolation."
—Joel Chandler Harris reporting of the Galveston Hurricane

What happened on that September day was that a hurricane with 130-mile-per-hour winds pushed a 15-foot-high storm surge across a coastal island whose highest elevation was eight feet. The destruction that followed—3,600 buildings totally destroyed, thousands more damaged—was unavoidable.

As for the horrific loss of life—6,000 to 7,000 people killed, making the Galveston hurricane the deadliest natural disaster in U.S. history—perhaps that, too, was unavoidable, given the circumstances in 1900.

Isaac Cline, riding up and down the beach, might have yelled, "Get off the island," except that, in this predawn of the transportation revolution, there was no way to get off the island. By early afternoon, the sand streets were under water, as was the single wagon bridge that led to the mainland. The three wooden railroad trestles had also been swept under, and, except for the trains, there was no motorized transportation in Galveston in 1900. According to Anna Peebler, of Galveston's Rosenberg Library, the automobile wouldn't make its appearance on the island until 1905.

Hurricane Andrew, equally devastating, caused few fatalities.

Florida state trooper oversees evacuation of the Florida Keys in advance of 2001's Hurricane Michelle.

Flash-forward 92 years, to dawn of August 24 on the southeastern tip of Florida, to Dade County and the townships of Homestead and Florida City, where ferocious Hurricane Andrew was about to wreak unimaginable havoc.

This was the big one. With damage assessments at $25 billion, Andrew was the costliest storm in U.S. history. According to a *Newsweek* story, Andrew "turned south Dade County into a zone of ruination that stretched on for miles and miles. Find the neighborhood and you couldn't find the street. Find the street and you couldn't find the house. Find the house and all you saw was debris." If anything at all remained, it was likely to be under guard. "MANNED AND ARMED, [the] graffiti said. YOU LOOT, WE SHOOT." (Aug. 14, 1992)

The toll of destruction included 126,000 houses and 9,000 mobile homes destroyed, 160,000 people left homeless in Dade County alone, the electric power grid wiped out in Homestead and Florida City, and 86,000 people without jobs.

Storm-related fatalities? Fifteen.

Yes, the people in South Florida certainly had better advance warning of the coming storm than the people in Galveston in 1900. They had emergency broadcasts on radio and television, and, instead of Isaac Cline riding the beach in his horse and buggy, they had policemen with bullhorns patrolling the neighborhoods. But most important, they had the Florida Turnpike and Interstate 95. They had the means to evacuate, and evacuate they did. Just from Dade County, 517,000 people made their way to safety.

According to Monroe County Emergency Management Director Billy Wagner, approximately 50,000 people evacuated from the Florida Keys in 24 hours. "We still had six hours before the storm hit," Wagner reported to a congressional committee, "and no traffic on the highway."

The transportation infrastructure was there when the people needed it, and it was instrumental in saving countless lives.

Stricken victims make their way from San Francisco after the 1906 earthquake.

California's reputation as a progressive state has been more than a century in the making, and her transportation infrastructure has always been an expression of that progressive mindset.

In 1895, California became one of the first states to pass legislation to form a state highway agency, and, in 1898, Los Angeles County oiled six miles of roads to lay the dust during the dry season—one of the earliest experiments in this technique. It's also worth noting that Lt. Col. Dwight Eisenhower, at the end of the famous cross-country military convoy of 1919 (discussed below) described the roads in California as "the best we had encountered."

As for San Francisco, in the year 1906, the sophisticated city by the bay was served by a sophisticated transportation system: well-designed and well-constructed streets; electric streetcars; the Southern Pacific Railroad, which had railheads both in the southern section of the city and across the bay in Oakland; and a small but growing number of automobiles.

When the great quake rocked the city on April 19 of that year, and as the three-day fire razed the city's center, the electric streetcars were of course immobilized. But a vast exodus from the stricken city was made possible by the Southern Pacific, whose headquarters and yards in the Mission District were saved from the flames.

U.S. Army troops from the Presidio were needed to maintain order in the destroyed city.

The Southern Pacific railroad helped in the monumental task of clearing the city of rubble.

Bird's-eye view looking west up Market Street.

What's more, unlike the ferrymen who gouged their customers mercilessly, the Southern Pacific ran its rescue and relief operation gratis. And the size of the operation was staggering: from the beginning of its effort, the railroad moved seventy passengers out of the city every minute in a shuttle service that eventually transported three hundred thousand people from the disaster area. It later offered free transportation for any refugee to any part of North America, meanwhile also shipping thirty-seven thousand

tons of relief supplies into the city at its own expense." (Thomas and Witts, 200-201)

And inside the city, approximately 200 privately owned automobiles were pressed into service, where they transported the injured to hospitals, carried troops and firemen where they were needed, conveyed urgent messages, and, perhaps most critically, carried the dynamite used to blow up buildings in order to create a firebreak. As the *San Francisco Chronicle* reported on April 29:

THAT THE AUTOMOBILE PLAYED AN ALL BUT INDISPENSABLE PART IN SAVING THE WESTERN PART OF SAN FRANCISCO, AND AT THE SAME TIME HAS PROVED INVALUABLE IN THE SERIOUS BUSINESS OF GOVERNING THE CITY THROUGH ITS GREATEST STRESS, IS CONCEDED BY EVERY MAN WHO HAS HAD HIS EYES OPEN DURING THE TEN DAYS OR SO THAT HAVE ELAPSED SINCE THE EARTHQUAKE. . . . MEN HIGH IN OFFICIAL SERVICE GO EVEN FURTHER AND SAY THAT BUT FOR THE AUTO IT WOULD NOT HAVE BEEN POSSIBLE TO SAVE EVEN A PORTION OF THE CITY OR TO TAKE CARE OF THE SICK OR TO PRESERVE A SEMBLANCE OF LAW AND ORDER.

Further testimony to the efficacy of motorized transportation—and to the existing transportation infrastructure—is seen in the fact that Walter C. White, head of White Motor Company, was able to organize a convoy of his motor trucks to bring emergency supplies up from Los Angeles to the stricken city. "In other words," writes John Rae, ". . . motorized highway transportation rapidly made its way into national life as something capable of meeting needs more effectively than they could be met otherwise, and in fact meeting needs that could not be met in any other way." (55-56)

Thanks in large measure to the city's transportation system—the evacuation and relief work of the Southern Pacific, the around-the-clock efforts of automobiles and drivers—the death toll from the tragedy remained at fewer than 500 fatalities, a remarkable figure given the size of the calamity.

The wind veered to the northeast, and lying in the cool sweet grass on the sloping hill, free from personal danger, we watched the fire burn forty blocks in the city's heart."
Henry Anderson Laffler
"My Sixty Sleepless Hours,"
McCLURE's, *July 1906*

The city was rebuilt quickly, as shown in this 1911 photograph.

The Great Fire roared down Market Street.

But the damage to the city had been done. And in 1906, with the heavy machinery of road building and repair still a dream of the future, simply clearing the streets of rubble seemed an overwhelming task. Every able-bodied man in the city was pressed into the service of removing debris by hand, brick by brick, and the machinery that did exist was ingeniously redesigned for the purpose. According to Saul and Denevi, "Railroad lines were stretched across the downtown area for clearing purposes. Steam and electric cranes lifted the twisted steel beams and dropped them upon flat cars. Huge mechanical devices for shoveling and loading were invented and put to work. Steam and electricity were used in ways never before imagined. But when it came down to cleaning up the mess, it was the men, along with horses and wagons, that had to get the job done." (136)

The earthquake of 1989 inflicted heavy damage to this section of Interstate-880 in Oakland.

ighty-three years later, when the Loma Prieta earthquake struck at five o'clock on the afternoon of October 17, Greg Bayol, then a public information officer with the California Department of Transportation (Caltrans), was at home watching the opening ceremonies of the third game of the World Series between San Francisco and Oakland.

"A lot of other people were already at home for the day, thankfully," says Bayol, now Caltrans' Chief of Public Affairs. "During normal rush-hour traffic, the disaster would have been much worse." It was bad enough. The 7.1 magnitude quake—the strongest to hit San Francisco since 1906—caused 62 fatalities and $6 billion in property damage. Damage to the transportation infrastructure totaled $1.8 billion—much of it on the San Francisco-Oakland Bay Bridge and, most tragically, the Cypress Street Viaduct on the Nimitz Freeway in Oakland, where 41 people lost their lives.

With his power out, Bayol walked down the street and found a neighbor watching a battery-operated television. When he saw footage of the Bay Bridge, he knew it was time to go back to the office.

"Our generator at work had failed," he recalls, "but we got emergency power up quickly and began assessing the damage. A cell-phone company gave us some phones—almost a novelty then—and our engineers worked throughout the night reporting on the status of the bridges. Thirteen of them ended up having to be closed to traffic at least temporarily.

"Once we finished the assessment," Bayol continues, "and knew what we were facing, we very quickly fell into a routine. Our district director, Burch Bachtold, told us to be creative, do anything we could, and if anybody offered help, take it—whatever it was. That really opened up the operation, because people did want to help—trucking companies, design firms. So yes, from inspection to assessment to work, we did move fast."

Highway construction workers quickly call in the heavy equipment.

A crane lifts a worker to the top of the destroyed Cypress Viaduct.

But it was enormously difficult work, under immense pressure. Bayol vividly remembers the major effort that took place 60 miles south of San Francisco in Santa Cruz, near the epicenter, where Route 17, a mountainous four-lane expressway, had been damaged and blocked by nearly 20 slides. "It all had to be cleared," he says, "and to get the trucks in, load them with the slide debris, and get them out again, we had to close the road to local traffic. We opened it at certain times, but these residents were trying to get contractors in, get the gas and electric people in to make sure their houses were safe. They were not happy. The *San Jose Mercury News* ran a picture with a guy holding up a sign saying, 'If you don't like what's going on, call this number.' Well, it was my number. Needless to say, we had to disconnect it. But we had that road fully repaired and reopened 30 days after the quake."

Lon Dugger, who was the area manager of Granite Construction Company at the time, oversaw the Route 17 operation, and he confirms Bayol's assessment. "Caltrans approached us immediately because they knew we had the resources and were equipped to take care of that kind of job. We went right to work for them on a 'force account' basis, our compensation based on their established rates. We put some of our best people on the job; in fact, I had my assistant manager there on a daily basis so that we could give Caltrans the best service possible."

Clearing the slide debris was only part of the job, recalls Dugger. "There were also huge fissures in the road itself, and we had people up there with grout pumps pumping grout into those holes. I have to say," Dugger declares, "Caltrans' ability to work with us, give us the right direction on an emergency basis, really made it possible for us to get in there and do what we had to do. We were up there right at a month, working 24 hours a day."

Amazingly, the Bay Bridge was also re-opened to traffic after 30 days. In the quake's aftermath, Caltrans promised to have the bridge operational within a month, and they made good on their promise. "Our engineers began that night to work on a design for repairs," recalls Bayol. "We found steel already fabricated to size up in Marin County and brought it down. We found a crane already on the bay ready to go to work."

Those crane operators and the engineers from Rigging International had the precarious job of first stabilizing and then gently removing the damaged sections of the bridge. As Rigging's Danny McLeod told the *Chronicle*, "The toughest part is to make sure these disturbed sections don't get away from us and fall on through. This is one helluva job. There's nothing in the book…. You just roll up your sleeves and get to work." (Oct. 21, 1989)

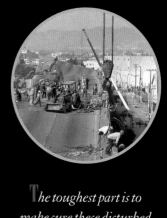

few days after the quake we kept having aftershocks, and we didn't know whether another big one was coming. The Oakland fire department headed that operation." Then came the tricky demolition work, where the parts of the viaduct not being actively demolished had to be carefully stabilized. "We brought in mountains of dirt," says Bayol, "and huge wooden support beams that we ordered from Oregon."

According to *Competing Against Time*, a report on the earthquake prepared by the State of California Office of Planning and Research, Caltrans had demolished and removed the standing portions of the viaduct and resurfaced the frontage roads by early January. (27)

To repair the broken levees along both the Pajaro River and the San Lorenzo River in Watsonville and Santa Cruz, Lon Dugger and Granite Construction again got the call. "Both of those levees had extensive damage," says Dugger, "and water was coming through. They actually had fissures big enough to walk through, six feet deep. The Army Corps of Engineers called us on a Thursday and said they needed a proposal by Monday. We got one to them, and within a matter of 24 hours, we were in a meeting and the contract was signed and we got to work. That was another around-the-clock project, and we were working on it concurrently. Again," adds Dugger, "the public agencies did what they had to do. Had their people there, and hired us on the spot, so we could get the work done."

On the Struze Slough Bridge in Santa Cruz County, 75 miles south of San Francisco, the deck collapsed and was impaled by the upright column, recalls Bayol, "but we got a contractor in

And hold on tight. Bayol adds that after waiting one day for the tide to rise enough for the 300-foot crane to reach the bridge, then the wind came up. "That was some dangerous work," he says.

Hazardous work was also underway to repair the Cypress Viaduct, California's first continuous double-deck freeway, two miles of which had collapsed. "The search and rescue effort itself was scary," says Bayol, "because during the first

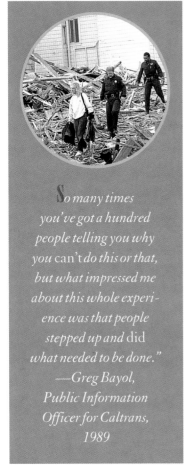

*So many times you've got a hundred people telling you why you can't do this or that, but what impressed me about this whole experience was that people stepped up and did what needed to be done."
—Greg Bayol,
Public Information Officer for Caltrans, 1989*

Policeman detours onlookers from another collapsed section of I-880.

there immediately and managed to get that bridge back in operation in about six weeks." And portions of the highways in the quake's path—Interstate 80 through Berkeley, 101 through San Mateo, the Bay Bridge toll plaza—"all these had to be all repaved," says Bayol, "and we got it done very quickly."

Bayol's job, of course, was to respond to the media, and it wasn't always pleasant. "The *Examiner* was particularly nasty," he says, but the press, in general, soon moved beyond its accusatory tone. "The thing was, we didn't waste time trying to defend ourselves. We repelled the hostile questions by saying, 'OK, our job right now is to fix this,' and we immediately met with professors from UC Berkeley and other experts to come up with a way to make the roads safe immediately. The way we attacked the problem helped bring the press around to our side of the matter."

Lon Dugger heard some of the criticism too, especially when Granite was called in to do some demolition work in the city of Santa Cruz. They had to take down and clear away some ten buildings that had been rocked by the quake and

were deemed unsafe. "People were protesting that they didn't want to lose this building or that building, but this was work that had to be done for public safety. Those buildings had been red-tagged, and they were dangerous."

Of course, the construction industry's work continued after the tremendous push of the first several weeks. *Competing Against Time* reports that although only 91 of the 1,900 state bridges in the area were damaged by the quake, Caltrans has retrofitted virtually all of them to make them more quake-resistant. And as Bayol observes, "We have come to understand, and the public needs to understand, that these earthquakes are our laboratory. We continue to learn very important truths about earthquake-resistant construction. We've also learned that the San Andreas fault is not the only one to be concerned about. Another major one, the Hayward, runs right beneath my office—and on into the center of Oakland."

But that tremendous response to emergency, those first days and weeks of difficult and sometimes dangerous repair work—those make for indelible memories.

"So many times," says Bayol, "you've got a hundred people telling you why you *can't* do this or that, but what impressed me about this whole experience was that people stepped up and *did* what needed to be done."

As Lon Dugger recalls, "This was a great opportunity for the public and private sector to work together and show what we can achieve when the chips are down. Yes, sometimes public agencies come under criticism for being so slow to act. Yes, sometimes, under normal circumstances, people can study something to death. But in this case Caltrans cut right through the red tape and let us get the job done. I have to say it was a great experience working with them under such circumstances."

Clearly, the whole operation was *can do*, and its remarkable success remains a tribute to the engineers, contractors, maintenance crews, and managing officials whose job it is to keep our transportation infrastructure the safest and best in the world.

The boost to national road improvement intended by the Federal-aid Road Act of 1916 was largely thwarted by America's entry into World War I. But at the same time, that war was vividly demonstrating the military and strategic importance of dependable roadways. Phil Patton notes that the railroads, during the war, "fell victim to a version of gridlock. The system virtually collapsed. Freight rotted on platforms. Lack of coordination between competing lines, facilities that had been allowed to deteriorate, and bad management caused either by complacency or by government regulation left vital war materiel languishing in depots.... By the end of the war, the long-term military uses of highways as an alternative to rail were on the mind of the newly returned General Pershing, [who] dispatched in 1919 a cross-country convoy." (80-81)

Most highway historians have paused to comment on this remarkable convoy—if only to note the fact that one of the officers who volunteered for the duty was Lt. Col. Dwight D. Eisenhower, who 25 years later would push for legislation authorizing the creation of the Interstate highway system. But it was a significant undertaking in its own right, an urgently needed test of both the country's roads and the military's mobile machinery. As Eisenhower himself recalled, "In those days, we were not sure it could be accomplished at all. Nothing of the sort had ever been attempted." (157)

On July 7, 1919, the motorized column of 79 vehicles departed from Milepost Zero in Washington, heading for San Francisco along the Lincoln Highway, with 260 enlisted men and 35 officers. "The trip would dramatize," Eisenhower wrote in his memoirs, "the need for better main highways. The use of Army vehicles of almost all types would offer an opportunity for

comparative tests. And many Americans would be able to see samples of equipment used in the war just concluded; even a small Renault tank was to be carried along." (157)

In fact, both the vehicles and the roads would leave much to be desired. After several speeches and the ceremonial dedication of the milepost marker, it was late morning when the convoy actually got underway—and early afternoon when the first breakdown occurred: "At 2:50 p.m. the Trailmobile Kitchen broke its coupling. A fan belt broke on a White Observation Car. And a Class B had to be towed into camp at the Frederick Fair Grounds with a broken magneto. The weather was fair and warm, the roads excellent," notes Eisenhower with some irony. "The convoy had traveled forty-six miles in seven and a quarter hours." (158)

If the machinery held up, the roads did not: "In some places, the heavy trucks broke through the surface of the road and we had to tow them out one by one, with the caterpillar tractor. Some days when we had counted on sixty or seventy or a hundred miles, we would do three or four." (159)

The U.S. Army's 1919 cross-country convoy takes a break near Big Springs, Nebraska.

Destination: San Francisco!

Road construction crew heads back to camp after a day of work on the Alaska Highway, 1942.

A THIRD OF A CENTURY LATER, AFTER SEEING THE AUTOBAHNS OF MODERN GERMANY AND KNOWING THE ASSET THOSE HIGHWAYS WERE TO THE GERMANS, I DECIDED, AS PRESIDENT, TO PUT AN EMPHASIS ON THIS KIND OF ROAD BUILDING. WHEN WE FINALLY SECURED THE NECESSARY CONGRESSIONAL APPROVAL, WE STARTED THE 41,000 MILES OF SUPER HIGHWAYS THAT ARE ALREADY PROVING THEIR WORTH.... THE OLD CONVOY HAD STARTED ME THINKING ABOUT GOOD, TWO-LANE HIGHWAYS, BUT GERMANY HAD MADE ME SEE THE WISDOM OF BROADER RIBBONS ACROSS THE LAND. (166-167)

If Eisenhower's peacetime cross-country convoy in 1919 advanced the argument that an adequate road system would strengthen the national defense, the bombing of Pearl Harbor in December 1941 elevated the argument to an imperative. With the U.S. Pacific Fleet crippled, the sea lanes between Alaska and the U.S. mainland were threatened, and no overland route existed.

President Roosevelt and Canadian prime minister Mackenzie King, considering their mutual borders and mutual interests, had already formed the Canadian-American Permanent Joint Board on Defense, and this board by 1940 was already overseeing the construction of the Northwest Staging Route, the chain of airbases stretching from Edmonton, in south-central Alberta, to Fairbanks. Now the board recommended that a military road be built, one that would link the airbases while at the same time serving as an all-weather land route to Alaska. The two countries agreed that Canada would supply the right of way and the U.S. would build the road.

The proposed route would cover 1,519 miles, winding from Dawson Creek, just north of Edmonton, across western Alberta, across northeastern British Columbia, through the southwestern part of the Yukon Territory, across the Alaskan border and on to Fairbanks. Timing was critical. The road had to be built in a hurry.

It took the column 56 days to make the trip—an average of roughly 50 miles a day—which no doubt sent a message not only to General Pershing but also to Congress, which would dramatically rewrite the Federal-aid Road Bill in 1921.

More important, Eisenhower's memories of the convoy, sharpened by his experience in Germany during World War II, would have a monumental impact on the development of America's transportation infrastructure:

military might moved swiftly into strategic position. Would Saddam's tanks, as feared, have rolled over Kuwait and on into Saudi Arabia had it not been for American troops immediately massing at the Saudi border? We can't know. But we do know he wasn't given the opportunity.

Lt. Col. Eisenhower, in 1919, could have scarcely imagined such a scenario.

Signallers aboard USS America direct fighter aircraft on night-lit flight deck after Desert Storm mission.

Aircraft in formation above USS America after news of ceasefire.

chapter five

INTRODUCTION

By Norman Y. Mineta, U.S. Secretary of Transportation

"The American Century," as the 20th century is sometimes called, might better be known as "The American People's Century." Whatever we accomplished as a nation, thousands of unsung Americans made it possible. ∽ That is certainly true of our transportation network. It is a product of great leaders, such as President Dwight D. Eisenhower, and celebrated visionaries, such as the Wright Brothers and Henry Ford. It is also a product of dedicated individuals who are little known today, such as American Road & Transportation Builders Association (ARTBA) founder, Horatio Earle, and Thomas H. MacDonald, who headed the U.S. Bureau of Public Roads from 1919 to 1953. ∽ But today's intermodal transportation network was also built by generations of American men and women—contractors, consultants, engineers, entrepreneurs, inventors, mechanics, environmental specialists, and others—who handled the nuts and bolts of building the greatest transportation network in history. They left their mark, if not their names, on our society by building a transportation network that supports economic opportunity, international competitiveness, and military preparedness, and by fostering the hopes and dreams of all Americans. ∽ In the 21st century, we will have great leaders, celebrated visionaries, and dedicated individuals, but also thousands of unsung men and women to create the transportation network that will serve as the foundation for what we as a nation, as a people, and as individuals will become. ∽ In recalling the unsung heroes of the 20th century, we thank them for the gift they have passed down, however temporarily, into the hands of yet another generation—our generation.

"Those who can, build. Those who can't, criticize."
—Robert Moses

Profiles in Leadership: The

According to the FHWA's *America's Highways, 1776-1976*, the appointment of Logan Waller Page, in 1905, as Director of the Office of Public Roads brought a new type of leader—the scientifically trained civil servant—into the highway movement. Page had been head of the road materials laboratory at Harvard and had studied at the famous French Laboratory of Bridges and Roads, which had helped make French roads the best in the world. His appointment announced the arrival of the highway engineer. ∽

Page and the men who followed him into the profession constitute an essential American type. As Phil Patton writes, our highway engineers "have been heralded for great, visible achievements, usually after spending the years creating those achievements in obscurity,

Highway and Bridge Builders

and have also taken the brunt of the

anti-technological waves that surge up regularly

from the heart of American pastoralism." These men were

among our "first true technocrats," with their firm belief in the

ability of research to solve the problems of the highways, and in the national

commitment to highways as a high cause. "They were men of great probity,"

says Patton, "dedicated, above all, to the highway 'system.'" (144)   They were

likely to have come from a rural background, where they could see first-hand the crying

need for improved roads. They might have seen one of Page's "good roads" trains come through

town, might have watched the construction of an "object lesson" road. They might have decided

that road improvement was important work. According to AASHTO's *The States and the Interstates*,

RED-JACKET
CRETE-BRIDGE
ICATION AUG 22-11

COPYRIGHT
1911
By
John R. Snow
MANKATO, MINN.

WILLIAMSBURG TUNNEL CONSTRUCTION

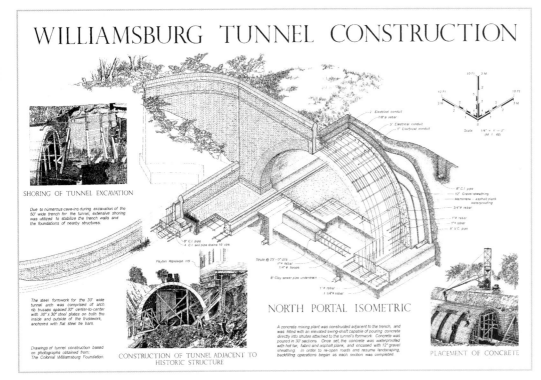

SHORING OF TUNNEL EXCAVATION

Due to numerous cave-ins during excavation of the 50' wide trench for the tunnel, extensive shoring was utilized to stabilize the trench walls and the foundations of nearby structures.

The steel formwork for the 30' wide tunnel arch was comprised of arch rib trusses spaced 30" center-to-center with 30" x 30" steel plates on both the inside and outside of the trusswork, anchored with flat steel tie bars.

Drawings of tunnel construction based on photographs obtained from: The Colonial Williamsburg Foundation.

CONSTRUCTION OF TUNNEL ADJACENT TO HISTORIC STRUCTURE

NORTH PORTAL ISOMETRIC

A concrete mixing plant was constructed adjacent to the trench, and was fitted with an elevated swing-shaft capable of pouring concrete directly into shutes attached to the tunnel's formwork. Concrete was poured in 30' sections. Once set, the concrete was waterproofed with hot tar, fabric and asphalt plank, and encased with 12" gravel sheathing. In order to re-open roads and resume landscaping, backfilling operations began as each section was completed.

PLACEMENT OF CONCRETE

Diagram of construction of the Williamsburg Tunnel on the Colonial Parkway in York County, Virginia.

New York City's highway system is already complex in 1946.

. . . MOST OF THESE [STATE ROAD ENGINEERS] MADE THEIR CAREER CHOICES EARLY ON AND PURSUED HIGHWAY WORK WITHOUT CONSIDERATION OF ALTERNATIVES. ENGINEERING, ESPECIALLY CIVIL ENGINEERING, APPEALED TO THEM, AND MANY CHOSE THIS FIELD DURING HIGH SCHOOL. . . .

ONCE EMPLOYED BY A STATE HIGHWAY DEPARTMENT, THEIR EMPHASIS WAS ON GETTING ROADS CONSTRUCTED TO SERVE TRAFFIC THAT WAS ALWAYS INCREASING IN WEIGHT, SIZE, SPEED, AND VOLUME. EVEN THOUGH INCREASES IN TRAFFIC VOLUME WERE LOWER IN THE RURAL DISTRICTS IN WHICH THESE MEN WORKED THAN IN THE LARGER CITIES, ENGINEERS REPORTED THAT LOCAL RESIDENTS NONETHELESS CELEBRATED THE COMPLETION OF EACH PROJECT AS PART OF AN OVERALL COMMUNITY DESIRE FOR GROWTH, PROSPERITY, AND PERSONAL MOBILITY. IN THIS WAY, YOUNG ENGINEERS INHERITED TWO TRADITIONAL AXIOMS THAT GUIDED HIGHWAY PROGRAMS — HIGHWAYS SERVED TRAFFIC AND THE PUBLIC SUPPORTED HIGHWAY CONSTRUCTION ENTHUSIASTICALLY. (40)

Over the century, our highway engineers have shared a vision and an ethic. They have seen themselves as at the service of the public interest, and they have served it well. "These engineers were singularly upright men," observes Patton. "The road program was free of the sort of corruption that attended the building of the transcontinental railroad. . . . This was due not to the honesty of politicians but to that of the engineers and technocrats." (146)

Engineers brought know-how to the transportation construction industry; they developed the technical expertise that made possible the monumental achievement that is our transportation infrastructure as a whole.

In that sense, engineers like Frank Turner, the first man to be profiled in the pages that follow, paved the way for the contributions to roadbuilding outlined in the profiles that follow his.

Frank Turner grew up in Texas farm country near Dallas—his father a railroad engineer, his grandfather a professor at a small college. Perhaps to tease his son, grandpa would tell Frank, "You know, boy, highways are the coming thing. The railroads have just about peaked. If I were a young man I would get into highways." (Patton, 143-44) In fact, Turner earned a civil engineering degree from Texas A&M and joined the Bureau of Public Roads (BPR), where he would spend his entire career.

Turner went straight from Texas A&M to a special training program the BPR had set up for its federal engineers. This was the "golden age of road research," and Turner was a devotee of studies and models, of advancing the science of the profession. But he was soon out of the classroom and out into the Bureau's most remote outposts. In 1942 he found himself plotting the route for the Alaska Highway from a bush plane. At the conclusion of the war, he was sent to supervise the re-

building of the Philippine road system. "It was a massive reconstruction project," Turner told *Roads and Bridges* magazine (June 1996). "Bridges of almost all sizes on the 7,000 islands that make up the Philippine archipelago were either severely damaged or destroyed by the Japanese."

His greatest work, however, lay before him. Back in Washington in 1954, Turner was appointed by President Eisenhower to be executive secretary of the Clay Committee, which was in charge of developing the plan for the Interstate highway system. The massive research and study effort behind the project could not have fallen into more capable hands: "We took off with those studies like a guy with a shiny new Mercedes," Turner recalled. "We really went to town with them." (Patton 149)

The Clay Committee report made two critical—and controversial—recommendations: that the Interstate system would tie into urban transportation networks, and that the new roads would be "free" rather than toll, to be paid for by the Highway Trust Fund. With the passage of the legislation in 1956, it would be up to Turner—Deputy Commissioner and Chief Engineer for Public Roads from 1957 to 1967 and then Director of Public Roads—to oversee the implementation of the greatest public-works project in U.S. history.

Turner (second from left)shown here at the signing of the Federal Aid Highway Act with President Eisenhower in 1954.

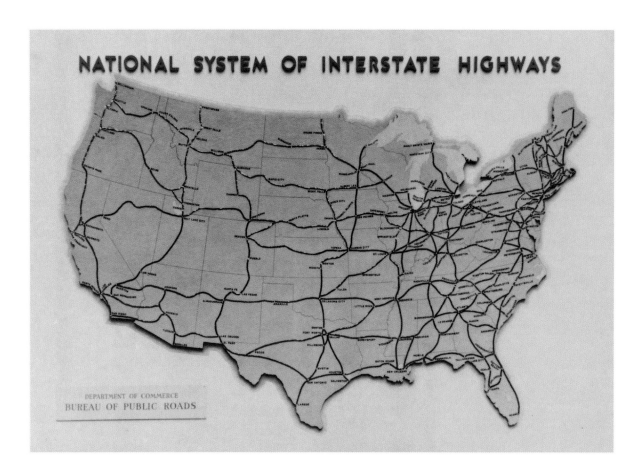

NATIONAL SYSTEM OF INTERSTATE HIGHWAYS

DEPARTMENT OF COMMERCE
BUREAU OF PUBLIC ROADS

I think people's opinions are pretty universal about the kind of man Frank Turner was.... He was a stalwart. He's someone we will seek to honor and emulate for years to come."
—Pete Ruane,
President, ARTBA

"Frank Turner was instrumental not only in the grand vision of the Interstate system," says Dean Carlson, formerly a top official at FHWA and now Secretary of the Kansas DOT, "but also in the nuts and bolts of getting the system created."

"He was the patron saint of Interstate highway development," adds John Yago, who served as Staff Director of the Senate Environmental and Public Works Committee and later as Director of Congressional Relations for ARTBA. "He was there before the inception of the system, he was there at its creation, and he was still there, at industry coalition meetings, long after it had been completed."

Turner went about the job with immense skill, quiet determination, and pride in being able to contribute to the nation's great work. Building highways had always been an expression of the public will, and Turner could not have foreseen that this attitude might change. But by the time he became Director in 1967 (his job title would change to Federal Highway Administrator in 1969), the first bugle calls of "the highway revolt" could be heard across the land.

Phil Patton observes that "As Federal Highway Administrator, Turner caught the brunt of the opposition to highways that developed in the late sixties and early seventies.... He was in the difficult position of presiding over the highway system at a time when thousands of people were displaced from their homes and thousands of small businesses destroyed to make way for the Interstates." (149) But Turner clearly saw his mission as larger than the plight of individuals, even if those individuals happened to be his own parents. Shortly after Turner's death in

1999, *U.S News & World Report* recalled the time Turner returned home to Fort Worth for a visit and found a stake in his parents' front yard. Interstate 35 was about to come in the front door. "Son, can't you do anything?" his mother asked.

"I can," he replied, "but I won't. You'll just have to move." (Dec. 27, 1999)

The incident exemplifies what friends and associates agree was Turner's cardinal virtue: absolute integrity. Patton reports that Turner was once offered a large chunk of stock in what was to become one of the largest motel chains in return for disclosing the locations of the interchanges on a planned new highway. He refused, of course. "Had I accepted the stock," he said afterward, "and I could have, I would be a millionaire today." (147)

During his 43-year career as a federal highway engineer, Turner received innumerable awards and commendations, including the inaugural Frank Turner Medal for Lifetime Achievement in Transportation, awarded by the Transportation Research Board in 1998. But certainly one of the most meaningful tributes to Turner came in 1994, when he was included by *American Heritage* magazine in its article "Agents of Change: Ten People Who Changed the Way You Live (and You've Never Heard of Any of Them)." While acknowledging the shifting public attitudes toward expressways in the latter part of the century, the article makes a telling point:

TODAY, ALTHOUGH [TURNER'S INTERSTATE SYSTEM] IS OFTEN WIDELY CRITICIZED AS DULL TO DRIVE AND AS A HOMOGENIZING FACTOR IN AMERICAN LIFE, THE INTERSTATES' BENEFITS ARE SO TAKEN FOR GRANTED AS TO BE BENEATH THE LEVEL OF CONSCIOUSNESS. AND THERE IS TESTIMONY TO THEIR POWER IN THE CONTEMPORARY METAPHOR FOR THE LATEST INFRASTRUCTURE DREAM: INFORMATION SUPERHIGHWAY. . . . DOESN'T ALL THE TALK OF THE WONDERS OF THE INTERNET POINT IMPLICITLY TO THE WONDERS OF THE INTERSTATES? (DEC. '94)

Not that honors and citations were of utmost importance to this modest, unassuming man, who never let the spotlight blind him to the task at hand. "What really counted for him in his retirement," writes Patton, "was the respect of the fraternity. AASHTO would call him out of retirement to address its annual conventions. Amid the workshops on such current topics as noise suppression walls and wildflower plantings, signs of the profession's need to placate a public once so enthusiastic about its work, Turner would recall the grand vision of the national highway system." (150)

Still, his industry found apt means to pay homage to Turner's distinguished career—and, specifically, to his belief in the value of research and technology in the improvement of road construction. The FHWA's research facility in McLean, Virginia, now bears his name: the Turner-Fairbank Highway Research Center. Moreover, a bust of Turner has been commissioned to be placed in front of the research center, with the inscription "Mr. Highways."

Deputy FHWA Administrator Lester Lamm, Frank Turner, Secretary of Transportation Elizabeth Dole, and Administrator Ray Barnhart at the Turner-Fairbank Center dedication.

THOMAS HARRIS MacDONALD: "THE CHIEF"

ANOTHER OF THE CENTURY'S TOWERING HIGHWAY MEN WAS THOMAS HARRIS MacDONALD, "THE CHIEF," WHOSE STERN COUNTENANCE ACCURATELY REFLECTED NOT ONLY HIS LEGENDARY FORMALITY BUT ALSO THE IRON WILL WITH WHICH HE RULED HIS DEPARTMENT.

THOMAS MacDONALD ATTENDED IOWA STATE COLLEGE OF AGRICULTURE AND MECHANICAL ARTS, AND, UPON GRADUATION IN 1904, IMMEDIATELY BECAME CHIEF ENGINEER OF THE NEWLY CREATED STATE HIGHWAY COMMISSION. WITHIN A DECADE, MacDONALD OVERSAW A COMMISSION STAFF OF 61 AND A BUDGET OF $15 MILLION.

For 34 years "the Chief" oversaw improvement of the nation's roads—like this Florida state route, circa 1929.

IN 1919, MacDONALD WAS NAMED "CHIEF" OF THE FEDERAL BUREAU OF PUBLIC ROADS, WHERE HE QUICKLY BECAME INSTRUMENTAL IN THE REWRITING OF THE FEDERAL-AID ROAD BILL IN 1921 — THE LEGISLATION THAT SPURRED NATIONAL ROADBUILDING IN EARNEST. "BY THE END OF THE DECADE,"

NOTES HISTORIAN TOM LEWIS, "THE CHIEF AND HIS BUREAU HAD BUILT OR RESURFACED OVER 90,000 MILES OF FEDERAL-AID HIGHWAYS AND WERE EXPENDING AN AVERAGE OF $78 MILLION EACH YEAR."

IN THE EARLY '40s, IN RESPONSE TO FDR'S PLAN FOR SIX "SUPERROADS" — THREE RUNNING EAST-WEST, THREE RUNNING NORTH-SOUTH — MacDONALD AND THE BPR ISSUED THE FAMOUS DOCUMENT, *Toll Roads and Free Roads*, WHICH ARGUED INSTEAD FOR AN "INTERREGIONAL" SYSTEM OF TOLL-FREE SUPERHIGHWAYS. THIS REPORT, AFTER WORLD WAR II HAD COME AND GONE, WOULD BECOME THE BLUEPRINT FOR THE INTERSTATE HIGHWAY SYSTEM.

APPOINTED BY WOODROW WILSON, MacDONALD WORKED UNDER A SUCCESSION OF DEMOCRATIC AND REPUBLICAN PRESIDENTS — HARDING, COOLIDGE, HOOVER, ROOSEVELT, AND TRUMAN — DURING A TENURE IN OFFICE OF 34 YEARS.

HE WAS 71 YEARS OLD — AND WIDOWED FOR 17 YEARS — WHEN HE RECEIVED WORD FROM THE SECRETARY OF COMMERCE THAT THE NEW PRESIDENT, EISENHOWER, DID NOT WISH TO RENEW HIS APPOINTMENT. LEWIS REPORTS THAT UPON HEARING THE NEWS, HE RETURNED TO HIS OFFICE AND ANNOUNCED TO HIS LONG-TIME SECRETARY, MISS FULLER, "I'VE JUST BEEN FIRED, SO WE MIGHT AS WELL GET MARRIED."

THE NEWLYWEDS LEFT WASHINGTON FOR COLLEGE STATION, TEXAS, WHERE MacDONALD WOULD SPEND HIS REMAINING YEARS AT THE MacDONALD HIGHWAY TRANSPORTATION CENTER AT TEXAS A&M UNIVERSITY.

President Harry S. Truman, left, greets John J. McCloy, center, high commissioner for Germany and Secretary of State Dean Acheson, right.

HARRY S. TRUMAN: 33RD U.S. PRESIDENT

HARRY TRUMAN BECAME A MEMBER OF THE AMERICAN ROAD BUILDERS ASSOCIATION COUNTY DIVISION DURING HIS DAYS AS JACKSON COUNTY "PRESIDING JUDGE" IN THE LATE 1920S. AFTER HE BECAME PRESIDENT, TRUMAN WAS AWARDED A LIFE MEMBERSHIP AND PRESENTED WITH A SILVER MEMBERSHIP CARD BY ARBA OFFICERS AND DIRECTORS.

Our 33rd president came by his devotion to roads honestly. His father was appointed "road overseer" for the south half of Washington Township in Jackson County, Missouri. "It was quite a job," recalls Harry Truman in his *Autobiography*. "He had to fix bridges and culverts, fill up mud holes and try to help everyone in the neighborhood get to and away from his farm in bad weather."

It was also his father's job to oversee the mandatory road work that was at that time a part of every able-bodied man's local tax burden. "A man could work three days on the road," Truman explains, "or he could pay the road overseer three dollars and let his work be done by proxy." If such an arrangement seems an invitation to graft, the invitation was declined by Mr. Truman: "It was my father's policy to actually work the roads for the money. . . . I was taught that the expenditure of public money is a public trust."

It was a lesson Truman kept in mind when, upon his father's death in 1914, he in turn was appointed overseer. "I served until the presiding judge became dissatisfied because I gave the county too much for the money." (36-37)

Given this intimate acquaintance with local road conditions, it's not surprising that when Truman himself was elected Jackson County's "presiding judge" in 1926, he immediately set about to rebuild the county's roadway network. He hired two engineers to conduct a thorough assessment, and their report concluded that the county faced "practically an impossible situation. . . . There are some 350 miles of what has been aptly termed 'pie crust' roads, clearly inadequate to stand the demands of modern traffic." (Daniels, 146)

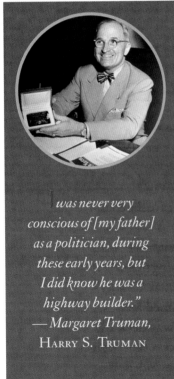

was never very conscious of [my father] as a politician, during these early years, but I did know he was a highway builder."
— *Margaret Truman,*
Harry S. Truman

Truman shows off the gold medal awarding him life membership in the AAA club of Missouri.

Based on the engineers' report, Truman proposed a $7 million bond issue, which, according to all the local politicians, had no chance of passing. Based on long previous experience, the voters had every right to assume that the money would be split between corrupt officials and their incompetent crony contractors. But Truman felt he could convince the voters otherwise, and, according to his daughter Margaret, "he launched a Truman-style campaign which took him once more into every corner of the county," promising that he would award contracts on a purely low-bid basis. When the votes were counted, "to the astonishment of all" the issue had been approved by a three-fourths majority. (Margaret Truman, 70-71)

Truman made good on his promise to award contracts with scrupulous honesty, spurning favoritism and earning the wrath of local political boss Tom Pendergast, who called him "the contrariest cuss in Missouri." But nobody could complain about the results. As Truman biographer Richard Lawrence Miller writes, "The completed road system was a wonder. Three hundred miles had been laid, over half in concrete. Professional engineers and honest contractors did the work so economically that some extra miles were constructed. And even then money was left over," most of which Truman transferred to his special road and bridge fund. (224) Farmers were especially gratified; as a result of the program, no farm in the county was more than two and a half miles from a hard-surface road.

Truman's interest in roads was not confined to those of his home county. The same year he was elected county judge, he also assumed the presidency of the National Old Trails Association, an office he held for many years. In this capacity, writes daughter Margaret, "he traveled extensively

VOTE "YES"
For the ROAD BOND ISSUE
NOV. 7th.
CANAL ST.
AT STATE
RAGE

His enthusiasm for this organization is apparent in a letter to his wife, Bess, written from Wheeling, West Virginia in 1927. "The National Old Trails Highway Association is a national organization indeed and in fact," he proclaims. "We are completely organized back here and have had our financial situation practically guaranteed." (*Dear Bess*, 327)

Back home in 1934, the county judge was about to step up on to the larger stage of national politics, and in his run for the U.S. Senate he logged countless campaign hours driving over the roads he had built a few years earlier. According to highway historian Tom Lewis, "Truman likened that campaign to being on a vacation trip. 'Fact is, I like roads,' said the candidate, 'I like to move.'" (164)

around Missouri and many other states, marking famous roads and urging local governments to see the value of their history as a tourist attraction." (68)

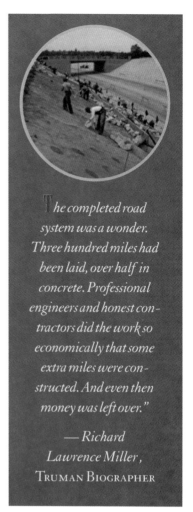

The completed road system was a wonder. Three hundred miles had been laid, over half in concrete. Professional engineers and honest contractors did the work so economically that some extra miles were constructed. And even then money was left over."

— *Richard Lawrence Miller*, TRUMAN BIOGRAPHER

Truman persuaded the local citizens to vote "yes" on his road bond issue.

Truman worked hard for the preservation of historic highways.

Road work was performed by mules and men in the early days of the Shepherd Construction Company.

It was by promising to build roads that politicians would get themselves elected back then, explains Clyde Jr. Everybody wanted a road leading to the county seat, so whoever was running for county commissioner would promise to spend some money on roads. "Of course, in the days before the gas tax, money for roads was hard to come by, so our job was to find out which counties had some revenue, get to know their commissioners, and let it be known that Mr. Shepherd could build roads if you needed them."

The Shepherd brothers agree that the basic procedure for "moving dirt" hasn't changed much over the years, but the equipment has changed radically. "We used wheelers back then," recalls Clyde Jr. "It was really a big pan on wheels, to scoop up the dirt, and you lowered the lip of the pan into the dirt, and the mules—the hook teams—pulled it forward. A plow team went first, to loosen up the dirt, and then the plow team might join up with the hook team to give you more mule power. To dump the wheeler once it was full, you'd try to get on a down-hill grade, then drop the tongue and get the men to flip it over.

"So basically, roadbuilding back then was mules and hard labor."

During those days, the Shepherd Company "outfit" consisted of roughly a hundred mules and a hundred laborers. "Your work crew might as well have been your family," says Harold Shepherd. "You fed them three meals a day."

"And when you were on a job," adds Dan Shepherd, "the first thing you did, after you fixed a place for the mules, was put up a cook shack so the food could be cooked and eaten in out of the weather. Everything else was tents, especially in

M y father had been in the dairy business," Clyde Shepherd Jr. explains. "He was about 30 years old when he bought a bunch of cows in Wisconsin. Beautiful animals, but when he got them home to Atlanta, it turned out they were infected with Bang's disease, and the health department condemned his herd and killed all his cows."

That was in 1918, a year that also saw World War I coming to an end and the automobile proliferating. "There weren't hardly any roads yet," says Clyde Jr., "but that was starting to change, and my father saw this as a propitious moment to get into the roadbuilding business."

Today Shepherd Construction Company is the oldest continuously-operating roadbuilding company in Georgia. It is still in the hands of Clyde Shepherd's three sons—Clyde Jr., Harold, and Dan—though the three brothers have bequeathed most of the day-to-day operations, as well as the executive titles, to their sons. Clyde Jr., now 85, is the chief repository of the company history, though his brothers also have stories to tell.

the summer time. In the winter the tents were shored up, and had woodstoves inside. No problem with firewood, of course, since you were always cutting down trees."

The Shepherd Construction Company shared in the prosperity of the '20s, but like everyone else, they felt the tremors of the crash in '29. "All of a sudden we had a company full of workmen and a yard full of mules," says Harold. "How my father managed to feed the men, the mules, and his family I will never understand."

With local work scarce during the depression, the Shepherd Company found jobs not just in Georgia but in neighboring states as well. And of course, the outfit served as its own transportation. As Dan says, "When we worked these out-of-town jobs, we would have to drive the team of mules to wherever we were going. We would just get out in the road and go. The workers would be riding the mules or else riding in a cart behind them. I mean, it was like a circus coming through town."

Although World War II slowed road-building across the nation, the Shepherds found ample work—and a way to contribute to the war effort—by building runways at the rapidly growing network of U.S. airbases. "Roosevelt was smart enough to start building up our armed forces long before Pearl Harbor," says Clyde Jr., "and we got the contract to build Camp Stewart down on the coast, the first anti-aircraft base in Georgia. The idea of war by air was in its infancy, but Germany's Luftwaffe was proving it could be done."

The Shepherd brothers figure they must have built ten airbases around the Southeast during the war—not to mention several in foreign countries. "We even built an airfield in the Azores," says Clyde Jr., "to help open up a more southern route across the Atlantic to Europe. The funny thing was, the nearest surplus labor was in Boston, and we had to go up there and hire 2,000 men, then get them on a ship down to the Azores. We built that base in the middle of the winter."

Building Camp Stewart in the early days of World War II.

Shepherd Construction built this airfield on Santa Maria Island in the Azores in 1944.

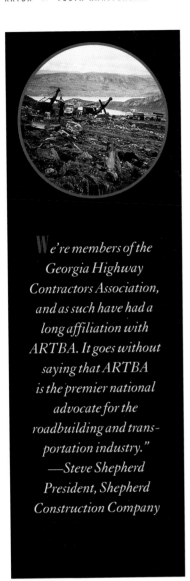

"We're members of the Georgia Highway Contractors Association, and as such have had a long affiliation with ARTBA. It goes without saying that ARTBA is the premier national advocate for the roadbuilding and transportation industry."
—Steve Shepherd
President, Shepherd Construction Company

Shepherd Construction was one of the major contractors to work on Interstate-20 in Georgia.

The Shepherds were well positioned for the arrival of the Interstate era beginning in the mid-'50s, and the company contributed greatly to the system in Georgia. "Our first big Interstate job was on I-20," recalls Harold, "but over the years we've worked on every one in the state. I guess our most impressive record was on I-95 along the Georgia Coast. Except for about 15 miles, we built that whole road, from the South Carolina line to the Florida line."

In today's regulatory environment, of course, it's a lot harder to

was originally planned—so everybody who travels that road travels six miles farther than he could have. Just imagine the economic—and environmental—impact of the millions of people who have driven that six extra miles. It's mind-boggling. Two hundred thousand cars a day go that extra mileage. The *Atlanta Journal-Constitution* had big coverage of how this wonderful person had saved Lake Allatoona."

Despite the vicissitudes of the industry, for more than 80 years Shepherd Construction has been building the roads that link the cities and towns of the Southeast—roads that the people have demanded and that have made those people's lives better. The Shepherd brothers' best guess is that they have built more than 10,000 miles of road in Georgia and neighboring states.

And with a fourth generation of Shepherd men poised to enter the business, the company's future looks bright.

Wheeler's Field in Libya, another of Shepherd's foreign projects.

build an expressway, or any other road, than it was forty years ago. The Shepherds have changed with the times, but, like many other veteran road-builders, they sometimes feel that they are being required to accommodate foolishness.

Clyde Jr. recalls, for example, the building of I-75 from Atlanta to Chattanooga, a route which passes over part of a large recreational lake called Allatoona. "First," says Clyde Jr., "keep in mind that Lake Allatoona is a man-made lake, nothing special or exciting about it. Some environmentalist type from a college up there decided that the highway shouldn't go across this lake, and he got a bunch of people to agree with him. As a result, the road is six miles longer than projected—it still crosses the lake, but not where it

Robert Moses at Brooklyn's Pierrepont Plaza Park in 1956.

Born in New Haven, Connecticut, in 1888, Robert Moses attended Yale, where he was elected to Phi Beta Kappa, and then Oxford. He was a brilliant student and a passionate idealist—committed, by the time he left Oxford, to a career in public service.

That career began in 1914, in New York City's Municipal Research Bureau, working for reform mayor John Purroy Mitchel. Guided by the idealist principle that jobs should be offered and promotions awarded based on merit, Moses devised a system by which every aspect of a city employer's performance could be assigned a numerical grade. Unfortunately, the city's Tammany Hall bosses found the young reformer and his system so odious that Moses was quickly out of a job. He absorbed the fundamental lesson: ideas are worthless without the power to implement them.

Jones Beach on Long Island, one of Moses' early triumphs.

Aerial view of Long Island Expressway system, circa 1946.

Moses began to accumulate the necessary power 10 years later when he became president of the Long Island State Park Commission and chairman of the State Council of Parks, titles he would hold onto—while adding others—for the next 40 years. With power came the shaping of his vision—the "public service" he wished to render and how to render it. His vision, to say the least, was expansive.

According to Moses' biographer Robert Caro, "No one in the nation seemed able to conceive of proposals—and methods of implementing them—equal to the scope and complexity of the problem posed by the need of urban masses for countryside parks and a convenient means of getting to them." Here was where Moses would find his early mission. In fact, a year before assuming the presidency of the Park Commission, he had already mapped out a system of state parks on Long Island that would cover forty thousand acres and would be linked together—and to New York City—by graceful parkways. Astonishingly, by 1929, writes Caro, "Moses had actually built the system he had dreamed of. . . . When Jones Beach, capstone of the system, opened, it opened to nationwide praise of a unanimity and enthusiasm not to be heard again for a public work until the completion of the Tennessee Valley Authority project a decade later." (10)

Indeed, the parkways Moses built during the '20s and '30s have become benchmarks in the history of highway building, and Moses' influence has been pervasive. As cultural historian Christopher Finch writes, "For the most part . . .

the story of the limited-access highway during [the '30s] is the story of the parkway proper, and the name most often associated with the proliferation of parkways was Robert Moses, among the most formidable highway builders of the twentieth century." (156)

Nassau County's Southern State Parkway, 1947.

Exemplifying Moses' best work during this period is the Meadowbrook Parkway, described by Finch as "a superb expression" of parkway design. The Meadowbrook, though, was but one of 16 parkways built by Robert Moses, totaling 416 miles. Moreover, the limited-access, multiple-lane, dual roadway would become elemental to Interstate design two decades later, and many young highway engineers came to the mansion on Long Island to study the work of Moses' favored design engineers—men like Gilmore Clarke, Michael Rapuano, and Clarence Combs. As Caro notes, they later assumed their places as "the road builders of America, the heads of state and city highway departments, key officials of the Federal Bureau of Public Roads. . . . Bertram D. Tallamy, chief administrative officer of the

Interstate Highway System during the 1950s and '60s, says that the principles on which the System was built were principles that Robert Moses taught him in a series of such private lectures in 1926." (11)

Of course, Moses himself went on to build quite a few expressways. His biographer states, in fact, that with the exception of the East River Drive, he built every major road in and around

New York City's Tri-Borough Bridge.

New York City. While he was at it, he built seven major bridges linking the boroughs that make up the metropolitan area—"immense structures," writes Caro, "some of them anchored by towers as tall as seventy-story buildings, supported by cables made up of enough wire to drop a noose around the earth." (6)

His expressway-building inside the city, however, was not so universally acclaimed.

Christopher Finch describes Moses as "an autocrat within the departments he controlled [who] never grew accustomed to the fact that the residents of densely populated urban areas were likely to object to having an expressway carved through the heart of still vital neighborhoods." (159)

One of his last major projects, the Long Island Expressway (LIE), is a case in point. Sixty-five years old when he conceived the project in 1954, he still wielded enough power to have the road approved by writing a simple letter to Governor Dewey explaining its necessity. But he heard the protests as the expressway was being built, particularly for his refusal to consider a rail line as part of the project. And even at the ribbon-cutting in 1962, writes *Newsday* journalist William Bunch, "Moses felt compelled to defend his vision, which equated the automobile with personal freedom and mobility, with the limitless opportunity for every Long Islander to drive to the office or the beach under his own horse-power, to the tick of his own personal clock. . . .

going a reevaluation today. But according to a *Newsday* survey, the people in Long Island and Queens are happily wedded to a lifestyle very much like the one that Robert Moses envisioned for them 35 years ago. "Nearly four out of every five respondents drove solo on the LIE. And 80 percent said they would not ride a bus to work, even if it could speed down an express lane and get them to work faster. And 77 percent said they would not join a carpool under those conditions. The master builder would have loved it." (*Newsday*, Dec. 4, 1988)

Ultimately, mythmakers and revisionists alike must stand in awe of the sheer size of Moses' creation. As Caro calculates, "[Including] only those public works that he personally conceived and completed, from first vision to ribbon cutting—Robert Moses built public works costing, in 1968 dollars, twenty-seven billion dollars. In terms of personal conception and completion, no other public official in the history of the United States built public works costing an amount even close to that figure. In those terms, Robert Moses was unquestionably America's most prolific physical creator. He was America's greatest builder." (10)

'There will be squawking no matter what we do,' [declared Moses]."

The protests continue to this day, as the LIE—along with the rest of Moses' expressway system—lies choked by gridlock morning and evening. Not surprisingly, Moses' achievements, borne out of his commitment to the private automobile at the expense of mass transit, are under-

Traffic crowds the Long Island Expressway, 1969.

Moses during construction of the New York World's Fair, 1963.

Stanard Lanford Sr. began his career in construction in 1923, working on Route 60 in Ansted, West Virginia. He was 17, and his job was to help drive blasting holes into rock with a steel-churn drill. One man held the drill's steel shaft steady while another pounded it into the rock with a sledgehammer. After each stroke, the man holding the drill twisted it a quarter turn. After a while, the men would change places, and eventually they would create a hole big enough to set a charge. Doing this for ten hours a day, Stan Sr. would later tell his sons, was the most back-breaking work he'd ever done. And it may have explained his eagerness to become a steam shovel operator.

Throughout the '20s and '30s, Stan Sr. continued to work as a shovel operator, often for Gilbert Construction, formed in 1926 by Enrico Vecellio, Littleton "Lit" Coleman, and Richard B. "Dick" Gay. His reputation was assured during a job in 1930 when, operating the company's new Bucyrus Erie gas-air shovel, he loaded 60 yards of material an hour during the summer season. Gilbert finished the job four months early, and the project was written up in *Excavating Engineer*.

In 1941 Dick Gay decided to leave Gilbert Construction and strike out on his own. His two founding partners in R. B. Gay & Company were Mann Gay and Stanard Lanford Sr. By this time Stan was married to Betty Compton, and the couple were raising a family: John Clayton "Jack" Lanford was already 10 years old, his little brother, Stan Jr., was seven, and their sister, Alice, was three.

The company stayed busy during the '40s, usually building, repairing, or straightening curves on railroad tracks. In fact, with both the C&O and the Norfolk & Western headquartered in Virginia, both Dick Gay and Stan Sr. settled their families in Roanoke. But in 1953 Dick Gay decided to retire from the business (his brother Mann having done so some years earlier), and a new company was born. Stan Sr. bought Dick out, and Ted Slater, as a reward for years of service to both Gilbert Construction and to R. B. Gay & Company, was given 10 percent of the new company, which would be called Lanford & Slater.

Meanwhile, Stan's efforts to dissuade his sons from following in his footsteps proved utterly futile. Both boys spent all the time they could on their father's job sites, and both worked construction jobs full-time during the summers of their high school and college years. Jack graduated from VMI in 1953 with a degree in civil engineering and immediately purchased 10 percent of the new company. Two years later, Stan Jr. earned the same degree from the University of Virginia and bought an equal stake.

The new company's first substantial road contract was for grading and paving three miles of Route 631 in Rockbridge County, Virginia. It proved a difficult job with very little profit in it. "Dad might have bid it pretty close," recalls Stan Jr., "because he knew the railroad work was finished and he wanted us to get a foothold with the DOT." Stan Sr.'s foresight was critical. "We learned a lot about working in Virginia and keeping the state road officials satisfied. Of course we've learned a lot more since then, and I think they're kind of glad to see us receive project awards."

Jack and Stan Lanford at ARTBA's Mid-Year Meeting, 1994.

Stan Sr. died of cancer in 1955, leaving majority ownership of the company—along with its fate—in the hands of his two sons. With money still tight during the late '50s, and facing what he considered an uncertain future, Ted Slater left the business in 1960. Jack and Stan renamed the company to Lanford Brothers Co. to emphasize its family orientation, and in 1962 they decided to specialize in bridge and culvert construction and repair. The move not only made sense, it brought the young company in close contact with English Construction and its Executive Vice President, Bill Owens. Thanks in large measure to subcontracted bridge work that the Lanfords did for English, the '60s became a watershed decade for the company, and its reputation grew along with its workload. "We could depend on the Lanford boys," remembers Owens. "If they quote they're going to do something, they're going to do it right, no question about it."

Daddy frequently reminded us that if we beat out more than eight other bidders, we had made a mistake."
— Jack Lanford

Another affiliation in the next decade proved equally important to the company's continued growth. Jack and Stan began doing bridge work for the J. F. Allen Company in West Virginia, where the projects were larger and more lucrative and where the company found room to expand. Ed Nuckols, one of Jack's original hires and a longtime project supervisor for the company, vividly recalls one of their West Virginia projects (this one for the state DOT) in 1976—repairing a bridge that spanned the Ohio River at Williamstown. "It was the most dangerous bridge repair we ever did and also the highest. We were working about 100 feet above the river for most of the time... People are still interested in how we repaired that bridge, because the fact is that one third of all the bridges in the country need repairs today."

If the Lanfords were fortunate to find work with companies like English and Allen,

those companies were fortunate to have men like Jack and Stan do their bridge work for them. John Allen, now president of J. F. Allen, acknowledges the mutual benefit: "Jack told me several times that they really needed that work and working with us helped them on the road to success. On the other hand, we also gained because they did quality work and were as trustworthy as anyone we could have ever found. They are just truly fine gentlemen."

Such testimonials—to their work and to their character—typify the high esteem in which the Lanfords are held by their colleagues in the industry.

With their bridge work steady and profitable, a different opportunity presented itself in 1985, which lured Jack back into the paving business. Doug

Dalton, now president of English Construction, wanted to purchase the venerable Adams Construction Company, which its principals were putting up for sale. The only problem was that Curtis English, whose approval had to be secured, discouraged the acquisition. He was afraid Dalton, his son-in-law, would be taking on too much. Dalton managed to change Mr. English's mind by persuading Jack Lanford to take on the management of Adams. As for Jack, the move suited him for professional and personal reasons. Not only did it offer a new challenge on a large scale, but it also left Lanford Brothers in the hands of Stan Jr., whose children were expressing an interest in joining the business.

Under Jack's leadership, Adams has become the second largest asphalt company in the state of Virginia. Meanwhile, according to Doug Dalton, Lanford Brothers with Stan Jr. at the helm has become the bridge reconstruction industry leader for the entire mid-Atlantic region.

cause: "There has never been enough money to provide for our road needs adequately. The interstate system is breaking up because these highways are overloaded and carrying more traffic than they were designed to carry.... The bottom line is that our industry associations are dedicated to assisting the development of quality roads and value-related construction for the needs of America."

Stan followed in Jack's footsteps, first serving with the VRTBA and then the Contractors' Division of ARTBA, presiding over that division in 1996. He joined ARTBA's executive board in 1997, and served as Chairman for 1999-2000. "I do not find any two brothers who have both held these offices in the past," says Stan. "I'm rather proud that both Jack and I can make that claim of service to our industry and citizens."

As for the future, Stan looks forward to cheering on the work of the next generation. Along with veteran managers Bob Milliron, Patrick McDaniel, and Al Soltis, three of Stan's children—Lynn Kirby, Margorie Cundiff, and son Ken—are firmly entrenched in the company. "There is an old adage about family businesses," says Stan. "The first generation makes the money, the second generation grows the money, and the third generation loses the money. But I believe that this company is well prepared for a third generation of success. I believe that when I walk out the door, our managers won't need me, which is the way I want it."

Huge safety nets protect Lanford Brothers workers on bridge construction project in Burnsville, West Virginia, 1977.

(inset, far left): Jack Lanford with Ed Nuckols on the job in Franklin County, West Virginia.

Heavy equipment: huge milling machine loaded on a low-boy trailer.

For Jack and Stan, success brings an obligation to give back to the industry that has nurtured them. As Jack says, "Stan and I began to understand that someone, or a group, has to lobby for highway needs in an intelligent way and press for funding."

Jack began by joining the Executive Committee of Virginia Road and Transportation Builders Association (VRTBA) in the early '70s and ultimately presiding over that organization in 1977. He then began to serve the industry association on the national level, culminating with his chairmanship of ARTBA in 1991.

It is service he is proud to have rendered. While conceding that lobbying Congress for road money carries the taint of "politics," Jack is quick to point to the urgency of the

chapter six

INTRODUCTION

By Terence M. O'Sullivan, General President
Laborers' International Union of North America (LIUNA)

This is the human story behind America's transportation system: people laboring in hot asphalt, wet concrete, and damp earth to build roads, bridges, and tunnels that are among the best in the world. Sadly, this is also a tragic story of loved ones killed in work zone accidents. The greatest tribute we can pay to the injured and the dead is not to increase their ranks; to drive safely, work safely, and protect those who labor on our behalf. ∾ For more than one hundred years, organized labor has been a part of building and maintaining our transportation infrastructure. We have trained safe, skilled, productive workers, given them a voice, and protected them on and off the job. We have worked with business and government to ensure the highest quality projects are produced safely, effectively, and on time. And as our communities grow, we remain committed to building the safe, efficient, and clean transportation system that our future demands. ∾ Read this chapter and learn about the human side of the roads and highways you travel every day. Think about the men and women who built them, and who sacrificed to benefit us all. We look forward to continuing to work with industry and government to strengthen and maintain the transportation system that is the envy of the world.

"The labor below is always attended with a certain amount of risk to life and health,
and those who face it daily are therefore deserving of more than ordinary credit."
—Washington Roebling
(in reference to the workers who suffered on the Brooklyn Bridge construction)

The Men and Women Who Built

Everyone is aware that people lose their lives on America's highways every day. Highway fatality statistics are widely published, and "predicted death tolls" are aired in advance of every holiday weekend. ❧ What most people don't realize, however, is that the work performed by the U.S. transportation construction industry over the past 50 years has *prevented* countless deaths from occurring and has, in fact, made our roads safer. According to Dr. William R. Buechner, vice president of Economics and Research at the American Road & Transportation Builders Association, the building of the Interstate system and the increasing use of these highways—in terms of total vehicle miles traveled—have had an enormous, positive impact on highway safety. ❧ Dr. Buechner reports that in the early 1950s, prior to the Interstate era, there were 7.0 fatal crashes for each 100 million vehicle miles traveled (VMT) in the United States. According to U.S. DOT figures, however, by 1996 the number had dropped to 1.7 fatal crashes per 100 million VMT.

America's Transportation Network

When we factor in U.S. DOT data

showing that the lowest fatality rate is on interstates

and other limited access freeways (the highest rates are on local

and two-lane rural roads), it becomes clear that the expansion of travel

on our modern, state-of-the-art expressway system is saving lives.

How many? Dr. Buechner calculates it this way: Though the total annual number

of highway deaths has grown from an average of around 35,000 in the early 1950s to

approximately 42,000 today, vehicle miles traveled over that period have increased *five fold*,

from 500 billion to an estimated 2.5 trillion annually. Based on that figure, if people were still

dying today on the nation's roads at the same rate as in the early '50s, the number of fatalities would

be more than 165,000 per year. "Thus it can be postulated," Dr. Buechner concluded, "that the total lives

'saved' over the past 40 years, versus the expected rate, is just over two million." Such a figure

is staggering, but it comes with its own cost. If the relative safety of today's highway system goes largely

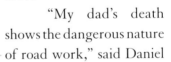

Construction worker sending up sparks on New Hampshire Route 101-A.

Daniel Doyle, Jr.

unrecognized, so too does the tremendous sacrifice of those who have worked to build that system. Literally thousands of unsung heroes have died over the past century building America's transportation network for the benefit of the nation's motorists, and highway construction work today remains hazardous. We rightfully honor our men and women who have fought to defend our freedom and quality of life each year on Veterans Day and Memorial Day. So, too, should we honor the men and women who have given their own lives while building the world's safest and most efficient transportation infrastructure network—a network that has provided America with an unmatched mobility and quality of life.

Just such a tribute took place on April 9, 2001, when a young man from Michigan joined the nation's traffic safety leaders at a ceremony at the National Mall in Washington, D.C., in commemoration of the 868 people—workers and motorists—who lost their lives during 1999 in roadway construction zone accidents. The young man's father, an employee of the Michigan Department of Transportation, had been killed while working on a bridge project in August 1997.

"My dad's death shows the dangerous nature of road work," said Daniel Doyle, Jr. "He was performing a great service that often goes unnoticed by the public. He worked hard to improve the roads and bridges for the residents of Michigan." Daniel Jr., now attending Lake Superior State University as one of the inaugural recipients of ARTBA's Highway Worker Memorial Scholarship Program, declared that he was proud to come to Washington to represent "the heroes in the roadway construction industry who have fallen in the line of duty," and that he was happy to do his part "to see that someone else does not lose their father, mother, brother, or sister."

But Daniel also made clear his wish that the opportunity had never arisen in the first place. "I would gladly trade innumerable trips to Washington—or anywhere in the world—in exchange for having my father back."

Building our transportation infrastructure has, from the beginning, been difficult work—hard physical labor, under conditions ranging from merely uncomfortable to extremely hazardous, and with the use of always newer and more complex technologies. Yet part of the distinction of the transportation construction industry is that historically it has offered such work—not just "honest" work, but important work, work to take pride in—to millions of men and women.

In terms of the scope of the human labor involved, few episodes in our history can match the building of the transcontinental railroad during the 1860s. Because of a scarcity of labor during a decade ravaged by the Civil War, the Union Pacific hired thousands of Irish immigrants to build the railroad west from Omaha, while the Central Pacific imported more than 10,000 Chinese workers to build east from Sacramento. From conception to completion, the massive project required six years, under the most extreme conditions.

In the winter of 1866-67, for example, as the Central Pacific was trying to cross the High Sierra, more than 44 feet of snow fell. As historian Wesley Griswold records, "The Chinese lived almost entirely out of sight of the sky that winter. Their shacks were largely buried in the snow. They walked to and from work through snow tunnels. They endured long,

Nevada campsite of Chinese crew at work on the transcontinental railroad.

grueling shifts underground in a dim, dank world of smoky lights, ear-ringing explosives, and choking dust.... Twenty of them were killed in a single snow slide.... Many other individuals simply disappeared, caught in small avalanches." (191)

Workers on the Union Pacific line, meanwhile, in addition to the perils imposed by the weather, had to endure sporadic but deadly Indian attacks. On a single day in 1867, Indian warriors killed and scalped five workers in a section gang near Overton, North Dakota, while another war party attacked and killed four graders working 100 miles further west. In perhaps the most chilling incident of all, on August 6, 1867, a party of Cheyennes looted and burned a freight train at the Plum Creek station in Nebraska, killing seven Union Pacific workers. (215ff.)

Historians can only approximate that "hundreds" of workers lost their lives in the huge undertaking. But once the last spike was driven at Promontory Summit, Utah, in 1869, fit tribute was paid them by General William Sherman, whose troops were called in for protection against the Indians. "All honor," he wrote to Union Pacific engineer Grenville Dodge, "to the thousands of brave fellows who have wrought out this glorious problem, spite of changes, storms, and even doubts of the incredulous, and all the obstacles you have now happily surmounted." (Griswold, 330)

Union Pacific work crew, 1868.

Construction engineers shake hands at the meeting of the rail lines at Promontory Summit, Utah, 1869.

The next decade, our country's massive bridges were beginning to be conceived and built, and the very concept of working conditions had to be redefined. In building his bridge over the Mississippi at St. Louis in the 1870s, James Eads borrowed the European idea of using a caisson—a chamber of compressed air—to construct the underwater foundations on which his huge steel arches would rest. Working inside the caisson was bad enough—hot, humid, backbreaking, and very risky—even without the disturbing effects some of the workers would experience after climbing back out of the deep underwater hole. The symptoms got worse—severe cramping, joint pain, headaches—the deeper the caisson sank, yet no doctor could diagnose it, or figure out why some workers suffered and others did not. By the time the bridge was finished, at least 16 men had died from what they only knew to call "caisson disease," and many more were crippled for life.

Nevertheless, at the very same time, a caisson three times as large as Eads' was descending to the bottom of the East River, on the Brooklyn side, inside of which a force of 264 men, with shovels, picks, crowbars, and sledgehammers would excavate the rocky bottom of the riverbed to a depth of 40 feet. In the harsh blue glare of calcium lamps, their grueling work went on around the clock, except for Sunday, in three shifts of eight hours each. Their pay was two dollars a day until, as the caisson sank deeper, the increasing difficulty and danger of the work necessitated a raise to $2.25. Many men quit, but for every one who did, a dozen—Irish, German and Italian immigrants—lined up to take his place.

As for the disease, few symptoms appeared during the work in the caisson on the Brooklyn side of the river. But on the New York side, where the caisson had to be sunk to bedrock at 80 feet, the effects of the disease became brutal. Three workers had died by the time Washington Roebling decided that the caisson was deep enough, and of the great number who had suffered he had these words to say: "The labor below

The east tower of the Brooklyn Bridge under construction in 1876.

is always attended with a certain amount of risk to life and health, and those who face it daily are therefore deserving of more than ordinary credit." (McCollough, 317)

Then, Roebling himself was stricken with the disease. For the next 10 years, while the towers rose to their full 276 feet, the huge cables strung from one side of the river to the other, and the vast project brought to completion, the chief engineer oversaw the work from the bedroom of his house in Brooklyn, with the devoted help of his wife, Emily.

Meanwhile, the transportation construction industry had found its own strong-voiced safety advocate: the American Road & Transportation Builders Association.

According to the association's 2001 Chairman John Wight, "For over three decades, ARTBA has led the industry in promoting work zone safety through its Transportation Safety Advisory Council, promotion and sponsorship of national work zone conferences and targeted safety spending. ARTBA and its member companies," Wight points out, "provide more financial support for targeted roadway work zone safety programs and initiatives than any other entity in the world. ARTBA member companies, such as those represented by the association's Traffic Safety Industry Division and Manufacturers Division . . . commit

GRACIAS!

Chinese in the West, Irishmen and Germans in the Northeast and Midwest—for more than 150 years, hardworking newcomers to America have played an important role in the building of our roads and highways. Today, especially in the South and Southwest, Hispanic Americans are carrying on that tradition. U.S. Census Bureau statistics suggest that an estimated 390,000 Hispanic Americans are part of the transportation construction industry workforce.

Dented barrels in Russellville, Arkansas, suggest the dangers of road construction work.

Nighttime highway work poses special dangers.

ARTBA and our partners have a long way to go until young men . . . like Daniel Doyle don't have to stand up at an event and say, 'I am here in the nation's Capital because my dad was killed in an accident that could have been prevented.'"
—*Pete Ruane, President of ARTBA*
Transportation Builder, April 2001

billions of dollars annually to safety research design and implementation." (*Transportation Builder*, May 2001)

Construction work zones are not going to go away. Today, much more repair and reconstruction is going on than new highway building. And during reconstruction, people are still going to use the road. It follows that the problem of worker safety will persist.

But the efforts of ARTBA and the transportation construction industry to ameliorate the problem will also persist—with increasing determination. ARTBA President Pete Ruane, though proud of the association's long record of work zone safety initiatives, concedes that the fight has not been won: "ARTBA and our partners have a long way to go

until young men . . . like Daniel Doyle don't have to stand up at an event and say, 'I am here in the Nation's Capital because my dad was killed in an accident that could have been prevented.'" (*Transportation Builder*, April 2001)

"Lives saved," of course, are more difficult to count than "lives lost." Nevertheless, it cannot be doubted that, thanks to the industry-wide response to the work zone safety issue, and thanks

in particular to ARTBA's leadership, the number of "lives saved" is steadily growing.

As John Wight says, "Imagine if your workspace was literally feet away from traffic speeding by you at 65 miles per hour." That's the environment facing highway workers today. Next time you drive through a roadway construction zone, salute those workers and be sure to "Give 'Em a Brake!"

Orange-striped barriers help protect these highway workers in downtown Pittsburgh

Lower Manhattan skyline with Brooklyn Bridge in foreground, at dusk, 1973.

chapter seven

INTRODUCTION

By Glen A. Barton, Chairman & CEO, Caterpillar Inc.

As we travel this country's great network of roads, few things are more familiar than the sight of yellow iron working tirelessly at roadside. And for more than 75 years, the yellow machines we've seen very likely came from Caterpillar. That's because beginning with the Better Roads Program of the 1920s and the Federal Aid Highway Act of 1956, and continuing today, Caterpillar has been there for every mile of this magnificent journey. ❧ It could be said that the greatest construction project in history—our interstate highway system—was and still is anchored by the heavy equipment industry. Caterpillar machines dot the landscape of America, clearing way for progress at every new turn. We've grown with the road building industry, and we like to think it has grown with us. Like no other industry, road building touches nearly every aspect of our business and economy. From mining the minerals and preparing the land needed for roads ... to paving the pathways trucks powered by our diesel engines travel carrying goods and services across the nation ... we've answered the call. ❧ It's difficult to imagine what this country would be like without the extensive network of highways, roads and interstates winding through our landscape. The transportation industry is just one example of why at Caterpillar, it's not just what we make, it's what we make possible. ❧ We're proud to have given our road building customers the power to change the world. And we look forward to the next 100 years of growth.

"The labor-saving machinery now manufactured for road building is just as effectual and necessary as the modern mower, self-binder and thresher."
—*Charles W. Wixom,* ARBA PICTORIAL HISTORY OF ROADBUILDING

Machines for a Big Job: How the Equipment

Gottlieb Daimler of Germany invented the internal combustion engine in 1886, and John Boyd Dunlop introduced the pneumatic rubber tire in 1889. These inventions would have an immediate impact on the rapid evolution of the automobile, but it would take a little longer for them to revolutionize the machinery that would build the roads for the automobiles to travel on. ∾ At the dawn of the new century, "good roads" became a national catch-phrase. The Office of Road Inquiry (ORI) had been established in 1893 under the direction of General Roy Stone, and in 1897 the ORI built its first object-lesson road—in front of the New Jersey Agricultural College and Experiment Station at New Brunswick. The road was 600 feet of macadam—a surface of clean, broken stones of a small, uniform size, with proper crowning and drainage, named after John Louden McAdam, the Scotsman who developed the technique. Another sign of the changing times was being waved by the bicycling organization known as the League of American

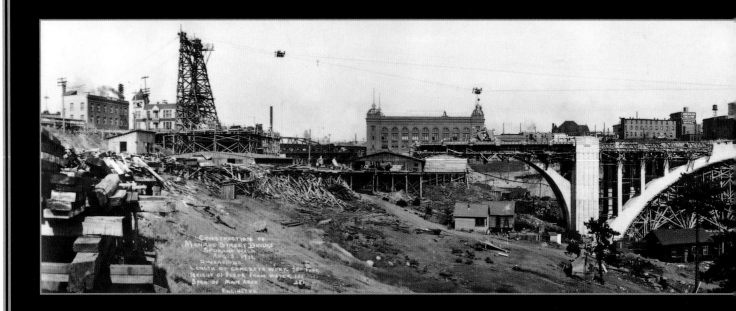

Manufacturing Industry Meets the Challenge

Wheelmen. According to transporta-

tion industry journalist Tom Kuennen, in 1899 the

Michigan chapter of the League—presided over by Horatio

Earle, who would soon found the American Road Makers—dropped

bicycle racing and adopted good roads as its main mission. Its first

International Good Roads Congress was held at Port Huron, Michigan, in July

1900. (*Transportation Builder*, Jan 2001, p. 11)   ∾   In its first complete census of

American roads, conducted in 1904, the ORI found 153,000 miles of surfaced roads in rural

areas, including 114,899 miles with gravel, sand-clay, shell, plank and other "low-tech" surfaces;

38,622 miles of water-bound macadam; and exactly 141 miles with surfaces superior to macadam.

But the point to bear in mind is that, in 1904, even the superior roads, like brick, were essentially made by

hand . . . or by machines—like Ely Whitney Blake's jaw rock crusher—powered by mules or men. For

example, the stones for a macadam road would be crushed to uniform size by a rock crusher—if one was

Sixteen horses pulling a road grader in North Dakota in the early days of the century.

The King drag in operation.

available—hauled to the right spot of the roadbed by wheelbarrow or dump carts, spread with rakes or the newly developed "spreading cart," and compacted by a horse-drawn roller.

Indeed, one of the most widely used "machines" of the day was the "King drag," developed in about 1905 by Ward King of Maitland, Missouri. According to the FHWA's *America's Highways*, a typical King drag was made from an eight-inch log split down the middle; the two halves were held about three feet apart by struts and the lower edge of the front log was equipped with a 1/4-inch steel cutting blade. A hitching chain attached to the front log

could be adjusted so that the drag would move earth to either side of the road as the device was pulled behind a team. (72)

Still, the ORI pushed hard for mechanization, even if it was mule-powered: "The labor-saving machinery now manufactured for road building is just as effectual and necessary as the modern mower, self-binder and thresher. Road graders and rollers are the modern inventions necessary to permanent and economical construction. Two men with two teams can build more road in a day with a grader and roller than 50 men can with picks and shovels, and do it more uniformly and more thoroughly." (Wixom, 56-57)

But the evolution of roadbuilding machinery would soon shift into high gear. Now the ORI began to speak of "building" roads, rather than "making" them, a sure reflection of the changes afoot. Helping set the pace, the American Road Makers in 1910 changed its name to the American Road Builders Association. Perhaps more significant, a year earlier at its annual meeting in Columbus, Ohio, the organization had

staged the inaugural "Road Show," an exposition of the latest in construction equipment and machinery. And in 1911, when ARBA met jointly with the American Good Roads Congress in Rochester, New York, the show drew 1,400 delegates from the United States and Canada, who flocked to the convention to admire the newest machinery and materials and to talk about the technical aspects of road building.

Meanwhile, a Californian by the name of Ben Holt was trying to figure out how to prevent huge, heavy, steam-powered agricultural machinery from bogging down in the soft soil of California's bottomlands. The answer, he decided, was tracks, and in 1904 he produced the first tractor-crawler. More important for the history of road construction, Holt was quick to realize that track-type machines had applications beyond farming and freighting. He built and operated a single-track machine with an adjustable blade suspended beneath it, and as early as 1909 this self-propelled grader was maintaining San Joaquin County roads for two dollars an hour.

New machines—and improving materials—were suddenly making a difference across the face of the nation. Portland cement, the binding material made from the ground-up combination of limestone and clay that had been developed in 19th-century England, was coming into use as a road surface concrete. And asphalt, a naturally-

occurring petroleum material, was beginning to be mixed with aggregate (sand or crushed stones) to form asphalt pavement. As Wixom reports, "In 1910, Portland cement concrete was used to pave about 20 miles of road, the next year 40 miles, in 1912, 250 miles more. By 1914, the total was 2,348 miles. . . . The amount of bituminous [or asphalt] concrete on rural roads was negligible in 1914, but in 1924 there were 9,700 miles of rural [asphalt] surfacing. (73-74) Tom Kuennen puts a dollar figure on the rapid pace of change: "The value of highway, road and street construction put in place in the United States in 1915 totaled $302 million. By 1920 it had more than doubled to $656 million." (*Transportation Builder*, Feb. 2001, p. 12)

1957 Road Show in Chicago.

Workers pouring asphalt from a hand-held pot on a Virginia road in 1916.

The "track-on-one-side, wheel-on-the-other" road machine developed by Holt in 1909.

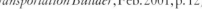

Already playing its part in the evolution of road-building machinery was Ingersoll-Rand, created in June 1905, by the merger of the Ingersoll-Sergeant Drill Company and Rand Drill Company, both of which were recognized leaders in the manufacture of power drills and air compressors. Ingersoll-Rand introduced its famous "jackhammer" in 1912, and by 1921 it had largely replaced the man-powered churn drill. Since the term soon became accepted popular coinage for that type of drill, regardless of the manufacturer, Ingersoll-Rand dropped one "m" from the name so that "Jackhamer" could become a registered trademark.

Ingersoll-Rand's original "Jackhamer."

Early Caterpillar motor grader at work in the California mountains.

Meanwhile, independent of the efforts of Ben Holt, the King drag had been supplemented by Adams leaning-wheel grader, which became motorized about 1920. Usually tractor drawn, this device brought a considerable advance in crowning and ditching operations. But its day, too, was about to pass. In April 1925, Ben Holt and his manufacturing competitor C. L. Best combined their operations into a new enterprise called the Caterpillar Tractor Company, and the roadbuilding equipment industry had a new driving force. Within five years the company would produce what it considered "the first true motor grader."

ROBERT LeTOURNEAU: PIONEER EARTH MOVER

Though born in Vermont, Robert LeTourneau found himself living in San Francisco, at the age of 18, when the great quake struck in 1906. The rebuilding of the city introduced him to the welding torch, a tool in which he saw vast possibilities. He pioneered the process of constructing machines entirely by welding rather than with rivets, and along the way he made welding a universally accepted industrial process.

In 1909, LeTourneau moved to Stockton, California, where he founded a dirt-moving business and began to build his own scrapers. In addition to the first all-welded scraper, he also invented the bulldozer blade that attached to the front of a tractor. In the '30s he began mounting his machines on huge rubber tires, bringing a new level of speed and mobility to the earth-moving process. At the same time he moved his plant to Peoria, Illinois, to be near the Caterpillar Tractor Company since many contractors used "Cats" to pull his scrapers.

His development of the self-propelled scraper-earthmover in the late '30s placed LeTourneau in the forefront of the earth-moving and heavy equipment industry just as World War II was beginning, and his company produced most of the earth-moving equipment (8,000 scrapers, 14,000 dozer blades, and other items) used by the Allies to build their critical roads and airstrips during the war.

After the war he moved his operations to Longview, Texas, and in 1953 sold his earth-moving business to Westinghouse Air Brake Corporation. Nevertheless, he continued to manufacture construction, roadbuilding, mining, and oil well-drilling equipment, and he was a leader in the design of mobile platforms for offshore drilling.

LeTourneau died in 1969, but "God's Businessman," as he came to be known, left a rich legacy. He poured 90 percent of his company stock into his LeTourneau Foundation, which sponsored Christian missions in South America and Africa, and which supported LeTourneau Technical Institute—now LeTourneau University—in Texas, along with several other educational institutions around the country.

Source: "LeTourneau, Robert Gilmour," by Ken Durham, from *The Handbook of Texas Online* (www.tsha.utexas.edu/handbook/online)

Originally called "auto patrols," these new machines were hailed by Caterpillar as a milestone in the road maintenance industry.

However, Holt and Best were not the only Californians who saw a bright future in the business of moving earth. In 1922, Robert LeTourneau produced his first all-welded scraper, and his ingenious design included electric motors that tilted the six-cubic-yard bowl for loading and unloading. A decade later, he devised a scraper with a cable-operated apron which enabled it to lift the bowl and carry the load.

1930 Road Show in Chicago's Coliseum.

Ingersoll-Rand air compressor from 1952 featuring pneumatic tires.

While manufacturers like Caterpillar and LeTourneau were steadily improving their road-building machinery, ARBA was providing them a showcase to display it. Thanks to the flow of federal highway dollars made available by the Federal-aid Road Bill of 1921, ARBA's Road Show became a major attraction for the road contractors and state and municipal road agency delegates attending ARBA's annual meetings. In fact, Tom Kuennen writes that by the mid-'20s, when the Road Show was taking place at Chicago's huge Coliseum, the manufacturing group known as the Highway Industries Association (HIA) had become a major participant, and in 1926 a record crowd of 15,000 attendees poured through the turnstiles to be brought up to date on the state of the art of road-building. The next year, significantly, the HIA—the forerunner of the Construction Industry Manufacturers Association (CIMA)—formally affiliated with ARBA as the association's "Manufacturing Division." (*Transportation Builder*, Mar. 2001, p. 19) CIMA became the Association of Equipment Manufacturers (AEM) on January 1, 2002.

At the end of the '20s and in the early '30s, tire technology caught up with roadbuilding machinery. As Wixom documents, tires on early automobiles were commonly inflated to pressures of 75 pounds per square inch, producing a hard ride and heavy roadway wear. Trucks, with solid rubber tires, inflicted even worse abuse on road surfaces. While Paul W. Litchfield, president of Goodyear Tire and Rubber, was working to develop pneumatic tires for trucks, his competitor Harvey Firestone was moving an important step further—developing a pneumatic tire for off-the-road equipment. When Firestone succeeded in introducing a large, low-pressure pneumatic tire about 1930, Robert LeTourneau, for one, was quick to recognize the impact of this improvement on the size, capacity and power of earth-moving and construction equipment. "We always had the horsepower," said LeTourneau. "What we needed was something to carry the load." (Wixom, 80)

As the science and mechanics of roadbuilding evolved, it became increasingly evident that the subsurface of a roadbed had even more to do with the road's load-bearing capacity than did the surface. Rollers had long been in use; in fact, the steam roller developed by Lemoine had been imported into this country before the turn of the century, and a steam-powered roller was in use on New York City streets as early as 1869. But now soil compaction technology was improved with the appearance of the subsurface, or "sheep's foot" roller, a device that in effect tamped the soil with rows of metal "feet" as it rolled along, pulled by a tractor or a team. Self-powered rollers of increasing sophistication were around the corner.

In the meantime, steam power itself had become obsolete. Steam shovels had been designed for use on railway beds as early as the mid-19th century, and for construction companies large enough to afford such equipment, they remained an important earth-moving tool through the turn of the century. Ben Holt's first crawler-tractors, too, were powered by steam, but even as he was placing these machines on the market, the advantages of gasoline power were becoming evident. He sold his first gas crawler in 1908 and never looked back.

In fact, he continued to look forward, and two decades later Caterpillar was the industry leader in the development of diesel power for road construction machinery. In 1931, according to company history, Caterpillar introduced the diesel engine to its tractor line and thereby pioneered the way for an industry that today runs almost entirely on diesel power. Full-scale production started in 1932, and the company's diesel production in 1933 exceeded that of the entire United States in the preceding year.

As the machinery used to prepare the roadbed—scrapers, graders, shovels, and rollers—continued to improve, so did specialized paving equipment. As Wixom notes in his *Pictorial History of Roadbuilding*, while concrete mixed on the job site had been used for road surfacing since about 1908, now a central mixing operation—both for asphalt and portland cement—began to be feasible with the development of dependable motor trucks to haul the mixed concrete to the site. The arrival of such

trucks also made possible the first double-drum paver—introduced in 1932 and capable of handling 27 cubic feet of mix in each drum. This doubling of paver capacity, in turn, spurred the development of paving trains. Huge drum mixers, fed cement and aggregate by truck, moved along on treads, pulling spreaders, levelers and finishers on rails. (112)

In the following decade, the Portland Cement Association's development of air-entrained concrete permitted longer hauls of slower-setting concrete to meet contractors' demands for ever larger volumes. Now multi-lane paving trains were laying down thousands of feet of concrete a day. (138) Asphalt paving was becoming equally mechanized, as asphalt plants increased in size and efficiency. The 1930s saw the introduction of asphalt pavers—road machines which ingested the heated asphalt-aggregate mix at one end and disgorged it in lane widths at the other, eliminating hand spreading on large jobs. Rollers and finishers followed to complete the paving job. The next step in asphalt

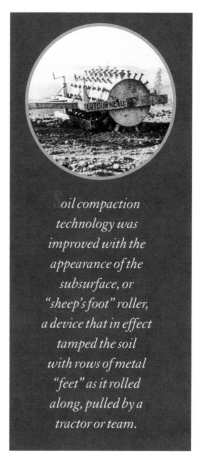

Soil compaction technology was improved with the appearance of the subsurface, or "sheep's foot" roller, a device that in effect tamped the soil with rows of metal "feet" as it rolled along, pulled by a tractor or team.

Paving train at work on Ohio's U.S. 40 in the 1940s.

My first recollection of ARTBA was when my father, Carl Stander, went up to the Road Show in Cleveland and bought a concrete mixer. That machine was used to pour a substantial amount of Ohio concrete pavement during the '30s."

— Richard R. "Dick" Stander, President (retired), Mansfield Asphalt Paving Co., and ARTBA Chairman, 1978

evolution was the on-site, self-propelled mixing plant, along with multi-lane spreaders for hot mix, and vibratory rollers which not only compacted sub-soils but provided base preparation and surface finishing as well.

Because of the onset of World War II, ARBA's Road Show had been suspended for eight years, but in Chicago in 1948 the show resumed with considerable fanfare. The attendance was a record 300,000 over nine days, including 330 exhibitors displaying some 650 pieces of equipment with an estimated total value of between $15 and $25 million. Among the newest wonders were a three-and-a-half-yard power shovel, a 24-ton crawler tractor, a 30-ton vibratory roller, a bottom-dump wagon capable of handling 20-yard loads, and a concrete paver boasting a 34-cubic-foot capacity. It was the greatest collection of construction equip-

ment ever, and for the first time actual demonstrations were possible, thanks to an outdoor display area.

At the conclusion of the 1948 Road Show the Construction Industry Manufacturers Association (CIMA) decided that, as a major participant in the show, it should assume administrative control. Nine years later, in 1957, the first exhibition was held under CIMA leadership—with ARBA as a principal sponsor. Thus, as Tom Kuennen points out, ARBA's original Road Show ultimately evolved into what we now know as CONEXPO/CON-AGG, one of the world's largest trade shows.

By the time CIMA put on its first show, however, one of the greatest advances in paving equipment of the 1950s was already well known. Actually conceived in 1946 by two Iowa highway engineers, James W. Johnson and Bert Myers, this dramatic innovation was the slip-form paver. This machine was designed to move along a prepared grade carrying its side forms with it, thus eliminating the need for hundreds or thousands of feet of steel forms and greatly speeding the paving process. Although the paver was used in Iowa in 1949, it was not made available generally until the mid-to-late 1950s—a time when, coincidentally, the ultimate need for it had arrived.

Chicago's Coliseum hosted ARBA's 1930 Road Show.

W ith the signing of the legislation to create the Interstate system in 1956, road-building machinery manufacturers saw that their time was at hand. And they immediately responded with bigger, faster, more powerful and more versatile equipment.

Case, for example, which had been in the roadbuilding machinery business ever since the launch of its 10-Ton Road Roller a half-century earlier, purchased the American Tractor Corporation in 1957 and in the spring of that year introduced the landmark Case Model 320, the industry's first factory-integrated tractor loader/backhoe. Three decades later, the Case loader/backhoe would listed by *Fortune* magazine among the "100 Products America Makes Best."

Caterpillar had already introduced its original front-end loader, the track-mounted No. 6 Shovel, in 1952, and the integrated tractor and

bucket concept continued to evolve as the size of the bucket increased dramatically. For many roadbuilding operations, tracks soon yielded to more maneuverable wheels, and Caterpillar marketed its first Wheel Loaders in 1959.

Caterpillar's Wheel Loader doing impressive quarry work.

Case's original Model 320 loader/backhoe.

Ingersoll-Rand's Crawlmaster drill at work on interstate project in 1959.

CMI's first production model AUTOGRADE, circa 1965.

The next year, Ingersoll-Rand introduced its Crawlmaster rock drill, touted as "combining the maneuverability of a light drill and the power of a heavy one." This machine was immediately put to extensive use for interstate highway construction in both Pennsylvania and New York.

Rising to the great challenge imposed by the Interstate building program, two other equipment manufacturers emerged in the mid-'60s, offering their own innovations to the massive project.

Operating on the belief that "the top of a road could be no better than the profile of the materials under it," Oklahoma City's Bill Swisher in 1964 set out to automate the road-grading process. The Interstate was coming to Oklahoma, and Swisher was concerned that, among its other inefficiencies, current road-grading practice could not accurately control tolerances. His goal was to design and build a machine that combined dual-lane subgrading, base material spreading, and finegrading—all in one, and all at tight tolerances. He called it the AUTOGRADE. With the

first production model scheduled for delivery the following year, in December of 1964 Swisher incorporated Construction Machinery Inc., which would come to be called CMI.

According to company history, the AUTOGRADE revolutionized the process for preparing grade and placing pavement. It was more productive, more efficient, less costly, *and* it controlled grade to within 1/8-inch tolerances—which improved contractors' ability to control the thickness of new pavement. By mid-1966, in perfect conjunction with the Interstate project, the Dual Lane AUTOGRADE was an accepted tool for the preparation of subgrade and base for concrete paving, and by the end of the decade every state in the U.S. required, by contract specification, the use of this type of machine. Before 1966 was over, CMI had also introduced its line of AUTOGRADE Slipform Pavers, thus consolidating its position as a key player in early Interstate paving.

Meanwhile, Harold Godbersen, a bridge contractor from Ida Grove, Iowa, had developed a double-oscillating screed finisher—a sophisticated leveling device specifically for concrete bridge deck finishing—in the early 1960s. To produce and market the new machine, Harold and his son, Gary, founded GOMACO as a division of Godbersen-Smith Construction in 1965. When state specifications in Iowa incorporated the use of the finisher, regional contractors began buying the machines, and the enterprise took wings.

The next year, GOMACO developed a cone drum or cylinder finisher to meet needs for skewability and finishing the wider bridge decks incorporated into expressway design. That machine evolved into today's C-450 concrete finisher, introduced in 1969, which made GOMACO a recognized name in the construction industry. Also in 1969, GOMACO expanded its offerings

crucial place in trenching and ditching operations, with capacities upwards of 10 cubic yards. Though these machines had been manufactured in the U.S. since the late-'40s, John Deere pioneered its all-hydraulic excavator in 1969. Caterpillar introduced its first hydraulic excavator, the Cat 225, in 1972 and by 1994 had become the world's leading producer.

The C-450 bridge deck finisher of 1969 established GOMACO's reputation in the construction industry.

This John Deere excavator represents the current product in a line first offered in 1969.

by developing an enlarged C-450 on tracks and combining it with a slipform paver.

As the '60s gave way to the '70s, the basic business of moving earth also continued to evolve. John Deere, a company whose brand had been established in agricultural machinery 150 years earlier, became a force in the roadbuilding industry when it introduced its first articulated graders in 1967. Caterpillar had introduced its first hydraulic scraper in 1962, and by the early 1970s self-propelled scrapers had mostly taken the place of tractor-drawn earth movers. Their payloads, too, were increasing astronomically. Early wheel tractor scrapers moved seven cubic yards; in the early '70s, their capacity had grown to 54 cubic yards. (For comparison's sake, the author's light pickup truck has a capacity of approximately four cubic yards.)

The late '60s also saw the emergence of the hydraulic excavator, which rapidly evolved to a

GOMACO's Commander III (GT-6300) pours and trims simultaneously.

Having expanded from bridge deck finishing into the slipform paving business, GOMACO in 1970 revolutionized curb and gutter construction with the development of the GT-6000 curb and gutter trimmer/slipformer. At the time, contractors were forming curb and gutter by hand with daily production of 200 feet. The GT-6000, with its advanced trim/pour concept allowing simultaneous pouring and trimming, allowed contractors to achieve 200 feet of production in a half hour. This machine evolved into 1974's COMMANDER III (GT-6300), capable of producing barrier and monolithic curbs, gutters, and sidewalks. (In fact, one of GOMACO's contractors, Shirley Concrete Company, used a COMMANDER III to slipform 16,625 lineal feet of curb and gutter in a single 11-hour day—the current world record.)

Now, gutters and curbs could be paved almost as fast as the roads themselves, the pouring of which was proceeding at breakneck speeds. Wixom reports that in 1950, the record paving day was 4,700 feet of pavement 11 feet wide. But by 1974, the American Concrete Pavement Association's "Mile-a-Day Club" included 70 contractors. The record day's work was 4.7 miles

BIG MACHINES FROM A BIG INDUSTRY

Needless to say, the few companies cited in this chapter represent but a small sampling of a very large industry. How large?

★ More than 840 establishments around the country manufacture heavy construction equipment.

★ The industry provides more than 81,000 jobs with a total payroll of just over $3 billion.

★ More than 160 companies in the industry have 100 or more employees, and the 12 largest have more than 1,000 employees.

★ More than 5,000 establishments in the U.S. deal in heavy construction equipment. Total employment in this corollary industry is approximately 84,000, with an annual payroll of $3.4 billion.

(source: Dr. William R. Buechner, "An Economic Analysis of the U.S. Transportation Construction Industry," 1999)

California construction company was using the machine to lay down between two and three miles of Interstate a day. Moreover, in 1970 CMI was joined by the Bid-Well Corporation, a major manufacturer of concrete finishing machines. According to company sources, since merging with CMI, Bid-Well volume has increased exponentially, and today CMI's Bid-Well Division is the nation's clear leader in the manufacture of Bridge Deck Pavers.

GOMACO's HW-165 paver was introduced in 1975.

of pavement 24 feet wide, placed on Interstate Route 80 in Idaho. "It was a long day," notes Wixom. "The contractor worked 23 hours." (165)

Both CMI and GOMACO were at the center of the pouring boom. As noted, CMI had launched into Interstate paving with the introduction in 1966 of its AUTOGRADE Slipform Paver, and before the end of the decade, a

In 1975, GOMACO expanded its paving line with the HW-165 paver, for secondary and street paving, then entered the interstate and mainline paving markets with its GP-2500 full-width slipform paver. In 1984 GOMACO introduced the GP-5000, capable of slipforming mainline and airport pavements in widths up to 50 feet, making it at that time the largest ever built for concrete paving.

leaves a clean, smooth surface suitable either for traffic or for bonding with a new surface. What's more, the machine salvages and recycles asphalt, which can then be reprocessed with new material and replaced on the road surface.

Given the fact that, according to National Asphalt Pavement Association (NAPA) figures, asphalt was being used to pave 65 percent of the Interstate system, it is not surprising that CMI had already entered the asphalt plant equipment manufacturing business while its ROTO-MILL was in development. In the '80s, as the sheer size of roadbuilding machinery began to give way to versatility and high-tech sophistication, CMI was building increasingly portable mixing plants with computer-controlled systems to meter out asphalt and aggregates to precise specifications. At the same time the company successfully designed a pollution-free asphalt recycling system.

While its machines were at work paving Interstate highways, moreover, CMI was anticipating the day—not far off—when new Interstate and other highway building would be substantially replaced by the job of highway repair and maintenance. The company developed a new process—Automated Pavement Profiling—and a new machine to accomplish it: the ROTO-MILL Pavement Profiler. The ROTO-MILL removes deteriorating road surfaces—reprofiling a full width of traffic lane at a single pass—and

In 1988, Ingersoll-Rand seized the opportunity to expand into the asphalt paving business with the acquisition of Fortress Allatt Limited's asphalt paving equipment division, and in 1990 the company furthered this expansion by acquiring ABG of Hameln, Germany, a manufacturer of equipment for roadbuilding and repair. With these acquisitions, Ingersoll-Rand's product line included milling machines to remove old surface, paving machines to apply the new surface, and vibratory compactors to make the new surface ready to use. Then, in 1995, the company's Road Machinery Division acquired Clark Equipment Company, an equipment manufacturer that included Blaw-Knox Construction Equipment. With the acquisition of Blaw-Knox, Ingersoll-Rand became the world's largest producer of asphalt road pavers.

CMI introduced Automated Pavement Profiling with its ROTO-MILL Pavement Profiler.

CMI designed its Hot Mix Asphalt Recycling System in 1978.

THAT'S SOME TRUCK!

Caterpillar introduced its first off-highway truck in 1962—the 769 with a 35-ton capacity. It seemed big at the time, but it was just a start.

Today's 797, a 360-ton (326-metric ton) is the largest of the Caterpillar line of mechanical drive trucks, as well as the largest truck ever built. According to a *New York Times* article, the truck is big enough to hold four blue whales, 217 taxi cabs, 1,200 grand pianos, or 23,490 Furbies.

And it has brains as well as brawn. Eight on-board computers control everything from tire traction and transmission shifting to air suspension seating.

In 1991 Caterpillar increased its presence in the road-paving arena with the purchase of Barber-Greene, world-renowned producer of paving equipment. The Barber-Greene Company had introduced the first practical asphalt paver in 1931, so its link to the paving industry rivals the longevity of Caterpillar. Today, Caterpillar offers a full state-of-the-art line of cold planers, reclaimers/stabilizers, asphalt pavers, and vibratory and pneumatic compactors.

Ingersoll-Rand rollers packing asphalt.

John Deere's extensive roadbuilding product line is typified by today's 1050C dozer.

GOMACO's GP-4000 features the In-the-Pan Dowel Bar Insertion (IDBI) system.

Case and John Deere also expanded their product lines during this period. In 1994 Case contracted with Sumitomo Heavy Industries of Japan to manufacture excavators for the North American market, and five years later the company merged with New Holland to become CNH Global, a world leader in construction and agricultural equipment. John Deere's construction machinery division, meanwhile, has grown to offer a product line that now includes 120 models of heavy equipment—including the latest in wheel loaders, backhoes, dozers, excavators, graders, scrapers and articulated dump trucks.

Also in the '90s, GOMACO patented its new two-lift paving system, which allowed two different concrete mix designs to be slipformed in a single pass. Another new innovation was GOMACO's In-the-Pan Dowel Bar Insertion (IDBI) system for two- or four-track pavers. GOMACO's IDBI system fits between the tracks of a standard paver and is the shortest in the industry from front to back. It is the only one available for a two-track paver providing consistent grade and slab depth while moving through vertical curves. The IDBI, says GOMACO, with its exclusive, computer-controlled, smart-cylinder technology is the most accurate dowel bar placing system in the world.

GOMACO'S product line today includes equipment that will slipform and/or finish concrete slabs from one foot to 150 feet wide: streets and highways, airport runways, curb and gutter, sidewalks and recreational trails, safety barriers, bridge parapets, and irrigation canals. Support equipment includes grade trimmers, concrete placers, concrete placer/spreaders and texturing and curing machines.

Also today, GOMACO is perfecting a stringless control system for their machines. The new technology allows GOMACO's slipform pavers, trimmers, and placer/spreaders to be controlled by an automated 3-D machine-control system and not by stringline. The 3-D control system is adaptable to any GOMACO Network or G-21 controller. The stringless system has already been used on projects around the world on airports, highway paving, sidewalk and curb and gutter.

Meanwhile CMI (now CMI TEREX, after its recent merger with Terex Corporation) has offered as its latest innovation a new series of variable width concrete slipform paving machines. This Hydraulic Variable Width (HVW) system, says the company, covers all concrete slipform paving markets from curb and gutter, to multi-lane expressways. According to Bill Swisher, the company founder who remains chairman emeritus today, "The flexibility and versatility of these new machines will forever change the way concrete paving jobs are organized and managed."

CMI's current pavers incorporate the Hydraulic Variable Width (HVW) system.

Case's state-of-the-art CX Excavator at work.

In fact, both GOMACO's stringless control system and CMI TEREX's HVW technology—with its premium upon the kind of versatility that can handle bigger or smaller jobs—typifies the "next generation" of transportation construction equipment. Across the manufacturing spectrum, as highway rehabilitation and maintenance to a considerable extent replace new road construction, sheer size and strength are sharing the workload with technological sophistication.

Case, for example, hails its new CX excavators as powerful "thinking machines," enhanced with onboard intelligence features. And Caterpillar, speaking more broadly, describes the coming generation of its equipment as "electro-x: in other words, electro-hydraulic, electro-mechanical, electro-pneumatic, electro-everything." Caterpillar, for example, describes the coming generation of its equipment as "electro-x: in other words, electro-hydraulic, electro-mechanical, electro-pneumatic, electro-everything." Electronics—in the form of computer microchips—will control all aspects of the machines' function. The company calls the high-tech computer operation now in development VIMS, or Vital Information Management System—described as an onboard doctor, policeman, data collector, mechanic, diagnostician, and production manager.

In fact, looking a few years down the road, Caterpillar foresees a day when its machinery will be largely autonomous—remote-controlled entirely by computer, tracked by global positioning systems, with no operators on board.

The companies highlighted here typify the innovative, can-do spirit of an industry that rose to the occasion of constructing "the greatest public works project in history" and that now faces the equally daunting task of safeguarding and enhancing America's transportation supremacy. As our transportation infrastructure evolves in the 21st century, featuring spectacular projects like Boston's "Big Dig" and ultra-modern intermodal facilities linking port, rail and highway—not to mention the crucial, if less glamorous, work of maintaining the systems already in place—we can be sure that our equipment manufacturers will continue to meet the challenge. The right machine will be ready ... whatever the job.

(Note: for the company-related materials used in this chapter, the author is indebted to *All in a Day's Work: Seventy-five Years of Caterpillar* (Gilbert C. Nolde, editor); *The Legend of Ingersoll-Rand*, by Jeffrey L. Rodengen; *CMI News* (20th Anniversary Issue, 1984; 30th Anniversary Issue, 1994; Winter 2001, and Fall 2001); and *GOMACO: The Worldwide Leader in Concrete Paving Technology*; as well as company promotional materials and web pages.)

chapter eight

INTRODUCTION

By Dr. T. Peter Ruane, ARTBA President & CEO

 The 20th century development of America's transportation infrastructure network—highway, bridges, ports, waterways, rail and subways—is one of the major reasons the United States is the world's remaining economic and military superpower. Cherished American values of freedom and mobility also thrived with construction of our infrastructure system. ∿ Despite these achievements, America at the dawn of the 21st century is in the middle of a growing infrastructure "capacity crisis." Since 1970, the U.S. population has increased 30 percent and vehicle miles traveled have increased more than 125 percent, but new highway capacity has only increased 6 percent. ∿ The capacity shortfall faces other modes as well. The number of airline passengers is expected to grow by more than 50 percent in the next decade. The volume of containerized cargo at our ports is expected to triple over the next two decades, and our railways and mass transit systems will also continue to grow. ∿ What is the future for America's transportation network? "Smart cars," self-financed "truck only" lanes, double-decker highways, more elevated trains and high-speed rail corridors, new tunnels under cities, even fixed guideways and "personal" flying vehicles—the possibilities are limitless. ∿ The future may be uncertain, but we do know two things. First, all levels of government will have to increase investment in modernizing and maintaining our transportation infrastructure and add capacity to meet the growing needs of a dynamic population and economy. Second, the men and women of the U.S. transportation construction industry will be there to pave the way.

"Congestion and inefficiency in transportation are not just inconvenient and aggravating,
but they are also a tax that burdens every business and every individual.
We have to find ways to lighten that load."
—*Norman Y. Mineta, U.S. Secretary of Transportation*

The Challenges Facing America's

The single goal of America's transportation construction industry has always been and will always be providing the fundamental freedom of mobility—mobility for business, mobility for pleasure, mobility for personal convenience, mobility in the face of emergency. For, as we have seen throughout the preceding chapters, mobility has in large measure driven the steady progress in our quality of life. ❧ Since the advent of the automobile age—a phenomenon that seems always to have outstripped any effort to measure and predict its impact—the transportation construction industry has had a huge job on its hands. For the first seven decades of the century—up through the initial surge of Interstate building in the 1960s—it was a job the industry attacked with relish and with confidence. The industry's hard work had the nation's blessing. Building a first-class transportation infrastructure for America was Job One. And everybody knew it. ❧ Today, unfortunately, that consensus is gone. To be sure, Americans still value mobility—now

Transportation Network Today

more than ever, since its palpable

advantages have been woven into the fabric of our

lives. But we are no longer of one mind as to how to achieve

and maintain it. Consequently, our mobility is threatened; the great

engine that drives American life appears to be sputtering. Our freeways are no

longer free, and millions of Americans are frustrated and befuddled by this perva-

sive and seemingly insoluble problem. ∿ How bad is it? A 1999 study conducted by

the highly respected Texas Transportation Institute (TTI) gets to the heart of the matter:

From 1982 to 1996, vehicle miles traveled (VMT) increased a spectacular 72 percent and U.S.

population rose 19 percent. But capacity, as measured in lane miles, increased a meager 6 percent

during those years. Broken down by year, while VMT was rising at the rate of 3.1 percent annually,

capacity was increasing by just 0.2 percent. ∿ The result of this tremendous imbalance between

demand and supply is predictable . . . and distressing. The TTI reports that in the 68 major urban areas it

ON THE LEFT A NEW ALLIANCE OF ENVIRON-
MENTALISTS, URBAN PLANNERS, AND ACA-
DEMICS BLAMED U.S. TRANSPORT POLICIES
FOR ENGENDERING SPRAWL, EXCESSIVE RE-
LIANCE ON CARS, AND THE RAPID DISAPPEAR-
ANCE OF GREEN SPACES. THIS COALITION HAS
ACHIEVED CONSIDERABLE SUCCESS IN SHIFT-
ING THE FOCUS OF TRANSPORTATION POLICY
FROM EXPANDING SUPPLY TO RESTRAINING
DEMAND. THEIR IDEA IS THAT GROWING CON-
GESTION CAN BE SOLVED BY GETTING PEOPLE
OUT OF CARS AND PLANES AND INTO BUSES AND
TRAINS. THIS HAS LED TO POLARIZING — AND
PARALYZING — BATTLES OVER POLICY. (50)

Similarly, explains Atkinson, the approach to the issue diverges along party lines. Preferring lower taxes to the higher quality of life inherent in transportation mobility, many Republicans would cut the gas tax, which would starve transportation infrastructure funding. Meanwhile, many Democrats, having cast their lot with the anti-car movement, have pushed for higher gas taxes but without firmly committing the money to road expansion.

Commuting in America author Alan Pisarski, writing in the same publication, sees highway expansion throttled not so much by partisan politics as by a combination of anti-highway activists and the myriad advocates of NIMBYism—the "not-in-my-back-yard" syndrome. The rhetoric generated by this coalition, says Pisarski, makes mass transit and cars competitors, even enemies, rather than teammates working toward a common end. Its central creed is "the widely held but utterly false belief that if we build enough mass transit, we will not need any more roads." The *reductio ad absurdum* of this argument is that if we make life miserable for 90 percent of travelers, some of them might switch modes of travel to work. "This is the use of congestion as a tool of public policy," writes Pisarski. "And it could not be a more self-defeating one. This is like solving a lack of food supply by passing out diet books." (*Blueprint*, 19)

surveyed, congestion has increased 213 percent since 1982. Today, the average U.S. driver spends 34 hours a year in traffic jams, and congestion now costs the U.S. economy more than $78 billion a year—more than three times the total in 1982. In Los Angeles alone, drivers sit stalled on highways 82 hours a year and spend more than $12 billion in wasted fuel and lost time.

The question, then, is why hasn't supply kept up with demand? Has the transportation construction industry walked off the job? In fact, the industry has not been allowed to do its job.

Writing in *Blueprint: Ideas for a New Century*, Robert D. Atkinson, vice president of the Progressive Policy Institute, explains how transportation policy has become a political football that both teams can punt but neither team can score with:

A NEW BREED OF CONSERVATIVES EMERGED THAT SAW IN PUBLIC TRANSPORT POLICIES THE WASTEFUL AND MARKET-DISTORTING HAND OF BIG GOVERNMENT. AND

TRANSPORTATION CONSTRUCTION: A RECORD OF REMARKABLE ACHIEVEMENT

★ The U.S. transportation construction industry has built 3.9 million miles of American roads and highways . . . more than 5,400 American airports . . . 200,000 miles of U.S. freight and passenger railroad track . . . 5,800 miles of urban mass transit lines with more than 2,300 stations . . . and 360 American ports.

★ Each year, the nation's transportation infrastructure handles more than 4 *trillion* miles of personal travel—an average of 15,000 miles per year per American—and more than $6 *trillion* worth of freight.

★ The U.S. highway and bridge infrastructure has an asset value of almost $1.4 *trillion*—about 10 times the asset value of all computers used in the United States.

★ Every $1 billion invested in transportation infrastructure generates more than $2 billion in U.S. economic activity.

★ The Federal Highway Program provides more financial resources for environmental and community enhancements than any other public or private effort. The total allocation for these purposes in 1998-99: more than $1.3 billion.

★ Public roads occupy less than one-half of 1 percent of the total U.S. land area. U.S. DOT data show total road capacity in the U.S. has only been increased 6 percent since 1982, despite a 19 percent increase in population and a 72 percent increase in the number of vehicle miles traveled.

Artist's rendering of the Dallas High Five project, the most expensive and complex road project in Texas history.

The fact is, however, that the forces marshaled against highway construction today are wider and more powerful than any particular interest group or coalition. Paradoxically, despite our demand for mobility and our frustration with the noxious effects of congestion, the public will seems to have succumbed to apathy and despair in the face of our transportation dilemma. As Pisarski wittily formulates the problem, "Too often today, we tend to see transportation only in terms of its negatives—the delays, the resources consumed, the lives lost, the pollution generated—to the point where our current goals for transportation can be met best by everyone's just staying home."

It seems to be a case of familiarity breeding contempt. The things we take for granted are much more likely to elicit criticism than praise. Why, even the trees in our suburban landscapes are more often abused for dropping their leaves on our lawns and gutters than hailed for the essential—but quiet—service they render. So it is that an important aspect of the challenge facing the transportation industry today involves the transformation—the rejuvenation—of the American mindset.

Transportation seems to be perpetually No. 11 on everyone's list of top 10 public issues. . . . In fact, transportation probably belongs no lower than No. 5 on our list of local public issues, after education, crime, health care, and jobs."
—*Alan Pisarski,*
Commuting in
America

In a speech delivered in the summer of 2000, Pisarski outlined a number of attitudes characteristic of this contemporary mindset, assumptions that have crept into the American approach to the transportation quandary. He calls them "certainties" that badly need to be reexamined:

★ *First: America's transportation system is pretty much complete.* A NATION THAT ADDS 25 MILLION PEOPLE EVERY DECADE, NOTES PISARSKI, WHOSE ECONOMY ADDS $4 TRILLION PER DECADE, THAT IS STILL A BEACON TO IMMIGRANTS FROM ALL OVER THE WORLD, CAN NEVER SAY THAT ITS TRANSPORTATION JOB IS DONE.

★ *Second: You can't build your way out of congestion.* YES, HE SAYS, YOU MUST OPERATE THE SYSTEM WELL; YES, YOU MUST MANAGE THE SYSTEM WELL. BUT YES, YOU MUST BUILD!

★ *Third: If you build more, it just fills up again.* SO-CALLED "INDUCED TRAVEL," A CONCEPT WHICH PARALYZES SO MANY TRANSPORTATION PLANNERS, IS WELCOMED BY PISARSKI. "THINK OF ALL THE 'INDUCED TRAVEL' WE WILL PRODUCE FROM GETTING PERSONAL VEHICLES INTO THE HANDS OF MINORITY POPULATIONS! WE SHOULD CELEBRATE IT, NOT CONDEMN IT," HE DECLARES.

★ *Fourth: We are a customer-driven agency.* "I AM AMUSED BY STATEMENTS SAYING WE ARE CUSTOMER-DRIVEN," PISARSKI SAYS. "CUSTOMER-DRIVEN ORGANIZATIONS MEET THEIR CUSTOMERS' NEEDS; THEY DO NOT PASS JUDGMENT ON THEM. TOO OFTEN OUR PLANNING HAS LOOKED LIKE PLOTTING AGAINST THE AMERICAN PEOPLE INSTEAD OF PLANNING FOR THEM."

Obviously, the "can-do" attitude that made it possible for America to build the world's greatest transportation system must be restored. Clear-sighted transportation thinkers across the spectrum agree that an efficient transportation system must be re-enshrined as a fundamental public good, and that the American will must be mobilized on behalf of mobility.

For example, the Democratic Leadership Council editorializes in its *Blueprint: Ideas for a New Century* that renewing our commitment to transportation is an urgent national mission:

RESTORING PERSONAL MOBILITY IS ESSENTIAL TO CONTINUING ECONOMIC GROWTH AND UPWARD MOBILITY. IT'S CRITICAL TO THE FORMER WELFARE RECIPIENT WHO CAN'T AFFORD TO BE LATE TO THAT FIRST JOB; THE TWO-EARNER FAMILY THAT MUST BALANCE WORK AND FAMILY NEEDS; THE FIRST-TIME HOMEOWNERS WHO SUDDENLY REALIZE THEY'VE 'BOUGHT' HOURS IN TRAFFIC JAMS; AND THE MILLIONS OF AMERICANS WHOSE HECTIC, DISJOINTED DAILY ROUTINES BEAR NO RESEMBLANCE WHATSOEVER TO THE PREDICTABLE RHYTHMS OF THE POST-WORLD WAR II ERA. MOBILITY DESERVES A MUCH HIGHER SPOT ON THE NATIONAL LIST OF PRIORITIES. IT'S A LEGITIMATE AND ESSENTIAL FUNCTION OF GOVERNMENT. (5)

During his confirmation hearing for the post of U.S. Transportation Secretary, Norman Y. Mineta endorsed the national commitment to mobility: "Transportation is vital to our national well-being, whether measured as economic growth, as international competitiveness, or as quality of life. Congestion and inefficiency in transportation are not just inconvenient and aggravating—though they certainly are that—but they are also a tax that burdens every business and every individual. We have to find ways to lighten that load."

But perhaps Pisarski put it best:

WE NEED TO STOP APOLOGIZING FOR TRANSPORTATION! WE ARE WORKING EFFECTIVELY TO AMELIORATE ITS NEGATIVE CONSEQUENCES. TRANSPORTATION HAS BECOME THE UNIVERSAL PUBLIC LEVER FOR ACCOMPLISHING ALMOST EVERY SOCIAL PURPOSE. WE EXPECT IT TO DO SO MUCH MORE THAN IN THE PAST. WE MUST DEDICATE OURSELVES TO ENHANCING THE BENEFITS OF TRANSPORTATION FOR THE NATION AND THE SOCIETY AND TO BROADEN THOSE BENEFITS FOR EVERYONE.

We need to remind those at work in DOT and FHWA every day of what a great service they perform and instill in them again the pride they should have in what they do. We need to teach young engineers, planners and technicians that building roads for America—expanding mobility for all Americans—is not something to be embarrassed about. It is a great public service.

The challenge facing the transportation industry today is to restore and maintain mobility for the American people and American business. The facts state the case bluntly: in a society where 90 percent of travel is by private automobile, we need more roads and highways, more capacity. No legislation will remove American citizens from behind the wheels of their automobiles. And the future, clearly, will bring more cars onto our roads, not fewer.

At the same time, our highways old and new must be made safer. According to the U.S. DOT's 1999 report to Congress:

★ POOR ROAD CONDITIONS OR OUTDATED ALIGNMENTS ARE A FACTOR IN AN ESTIMATED 15,000 ROAD-RELATED FATALITIES A YEAR (THAT'S 36 PERCENT OF THE ANNUAL HIGHWAY DEATH TOLL OF 42,000).

★ 28 PERCENT OF ALL ARTERIAL ROAD MILES IN THE U.S ARE IN POOR OR MEDIOCRE CONDITION. ON URBAN INTERSTATES, 36 PERCENT OF THE PAVEMENT MILEAGE IS POOR OR MEDIOCRE.

★ THROUGHOUT THE NATION, 172,572 BRIDGES — 30 PERCENT OF THE TOTAL — ARE EITHER STRUCTURALLY DEFICIENT OR FUNCTIONALLY OBSOLETE.

★ TRAFFIC ACCIDENTS ARE THE LEADING CAUSE OF DEATH AMONG AMERICANS FROM SIX TO 28 YEARS OLD, AND THE COST TO OUR SOCIETY IS MORE THAN $160 BILLION ANNUALLY.

The American Road & Transportation Builders Association, the National Safety Council, and many other organizations are working together to give highway safety a high policy priority, but certainly expanded highway capacity is part of the solution. Too many cars on too few miles of roadway will always spell danger.

And, of course, while expanding the nation's transportation infrastructure and enhancing the safety of our highways, the industry must remain sensitive to environmental concerns. The judicious development and use of mass transit will play a crucial role here, as will a number of technological advances—both automotive and roadway—already at hand. But the right balance must be struck between the larger public good of mobility and the sometimes narrow vision of special interests.

The Theodore Roosevelt Bridge, Washington, D.C.

Rush-hour traffic on the Los Angeles freeway.

The challenge of restoring America's mobility, however, extends far beyond the problems faced on the nation's highways. Our entire transportation infrastructure—including airports, mass transit systems, railways, ports and waterways—shares in the great challenge of mobility, the challenge of creating a transportation system that is seamlessly intermodal. In the broadest sense, transportation's problem is too many people and goods in too few places, with too little capacity to keep them moving. Adding capacity—throughout the various modes of travel—will have to be part of the solution.

Here is a brief look at the major components of our transportation infrastructure:

Air-traffic controllers at their instrument panels inside the tower at La Guardia Airport.

Subway train pulls into the Rhode Island station.

AIRPORTS.

The nation's airports are the source of almost as much frustration as our highways. Like ground travel, air travel has surged—to 650 million passengers a year—but the construction of airport infrastructure has not kept pace. The inevitable result: delays, cancellations, and congestion. According to *Newsweek*, during the year 2000 one out of every four flights ran late or didn't run at all, and some experts put the cost of delayed and canceled flights at $5 billion a year.

In its comprehensive cover article, *Newsweek* looks at the wide range of problems facing air transport today and offers seven suggestions for improvement. Significantly, its "Solution No. 1" is to "pour more concrete." The article points out that, largely because of intense local opposition (NIMBYism at its most vehement), only two major airports—Denver and Dallas/Ft. Worth—have been constructed in the United States since the start of jet service more than 40 years ago.

Conceding that any new airport near a major metropolitan area would have to clear literally impossible regulatory hurdles, *Newsweek* suggests the perhaps more tenable solution of adding runways at existing airports. In the last 10 years, according to the article, only six new runways have been built at the nation's 30 busiest airports—the airports that handle 70 percent of all air traffic. Experts say 25 new runways are needed now to ease congestion, and while some 15 are in the works, most won't be completed for several years. Indeed, approval for runway expansion is almost as difficult to obtain as permission to build a new airport. More than one commentator has pointed out that, thanks to the various levels of bureaucratic review, Memphis needed 16 years to complete a new runway, even though there was no local opposition to the project. (*Newsweek*, April 23, 2001)

Perhaps in the short term, there is much to be said for *Newsweek*'s Solution #2: "unclog the airways" by improving the management of airport demand. Dorothy Robyn, currently a guest scholar at the Brookings Institution, points out that, like highways, airports typically handle the bulk of their traffic during early morning and late afternoon peaks. She says that a restructuring of airport landing fees would provide incentive for users to shift flight activity to off-peak hours or to less congested airports—especially turbo-props and regional jets which predictably clog airports during peak hours. (*Blueprint: Ideas for New Century*, 36-37)

But the overwhelming need for expanded capacity must be confronted sooner or later. And given FAA forecasts that air traffic will rise another 50 percent in the coming decade, the sooner the better.

MASS TRANSIT.

If Americans seem to like the idea of mass transit more than they like the actual practice of it, that may be because mass transit has been unable to offer a compelling alternative for much work-related and personal travel.

Transportation analysts have noted that due to its fixed routes and schedules, public transit is never likely to meet all, or even a high percentage of, our daily transportation needs, particularly outside of dense population centers. Recent surveys show, for example, that work travel constitutes only 25 percent of total passenger mileage and that, overall, Americans use transit systems to make only about 2 percent of their trips. Yet broken out as a component of the transportation equation, that 2 percent reveals some impressive numbers.

According to the U.S. DOT's Bureau of Transportation Statistics, 382 million individual trips were made on commuter rail systems in 1998 (the most recent year for which figures are available), and these trips added up to 265 million passenger miles traveled. Motor bus transit systems, meanwhile, accounted for 5.4 billion individual trips, and 2.3 billion passenger miles traveled.

Altogether in 1998, passengers used transit for 8.7 billion trips and traveled 3.9 billion miles.

Clearly, transit is playing a vital role in meeting our mobility needs, especially in densely populated urban cores. What's more, several factors will make transit's role increasingly vital: population growth, the economy and convenience of transit travel, and the growing trend toward intermodality seen, for instance, in the conjunction of air and rail terminals. Will transit's infrastructure meet the growing demand? The Bureau of Transportation Statistics reports, somewhat ominously, that the average age of our commuter rail passenger coaches is 19.4 years. Capacity expansion, along with capital improvements, will certainly be necessary if our public transit system is to fulfill its potential as a prime mover of people.

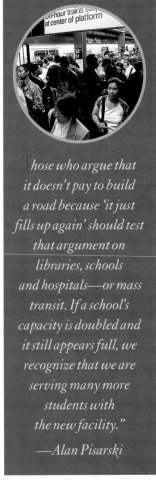

"Those who argue that it doesn't pay to build a road because 'it just fills up again' should test that argument on libraries, schools and hospitals—or mass transit. If a school's capacity is doubled and it still appears full, we recognize that we are serving many more students with the new facility."

—Alan Pisarski

Intercity freight yards support metropolitan commerce.

Containerized cargo gets a huge lift onto CSX rail car.

Our freight railroads directly contribute $13 billion a year to the U.S. economy in wages and benefits to more than 200,000 employees and billions more in purchases from suppliers. Today there are more than 1.3 million freight cars rolling over America's rails, with a carrying capacity of close to 128 million tons—an 18 percent increase since 1990.

But the growth rate in rail freight—and its implications for our transportation infrastructure needs—shows its true proportions when consideration is given to intermodal transport (defined by the American Association of Railroads [AAR] as the movement of trailers or containers by rail and at least one other mode of transportation). According to figures published by the AAR, intermodal traffic has grown from 3.1 million trailers or containers in 1980 to 8.8 million in 1998—an increase of almost 300 percent.

Some of this huge volume of traffic is now handled by such innovations as double-stack trains—i.e., with one container atop another—and so-called RoadRailers, which can travel on either rails or highways. But increased volume, whether double- or triple-stacked, inevitably means increased strain on the infrastructure. Today, freight railroads, a privately owned and deregulated industry, pay their own way—to the tune of $247 billion invested since 1980 in rail infrastructure maintenance and improvement. As rail freight continues to expand, so must this investment.

FREIGHT RAILROADS.

When we remind ourselves that an essential aspect of American mobility is the movement of goods and that, in fact, the movement of goods is the lifeblood of American business, another component of our transportation infrastructure—freight railways—comes sharply into view.

Those who think of railroads nostalgically (or who don't think of them at all), should consider that railroads carry—

★ MORE THAN 40 PERCENT OF THE NATION'S INTERCITY FREIGHT;

★ 70 PERCENT OF VEHICLES FROM DOMESTIC MANUFACTURERS;

★ 64 PERCENT OF THE NATION'S COAL, WHICH GENERATES 36 PERCENT OF OUR ELECTRICITY;

★ *and* 40 PERCENT OF THE NATION'S GRAIN.

PORTS AND WATERWAYS.

Of course, freight also moves over our nation's waterways, in steadily increasing amounts. Hugh O'Neill, formerly director of New York State's Office of Development Planning, calls America's ports "our gateways to economic growth," and figures from the Bureau of Transportation Statistics would seem to support his assertion. In 1960, approximately 1.1 billion tons of goods were transported via water; by 1998 that total had more than doubled to 2.3 billion tons.

What's more, during the past decade (1990-1999) the nation's busiest ports got much busier. At No. 1-ranked Southern Louisiana, freight tonnage increased 10.3 percent. At No. 2 Houston the increase jumped to 25.9 percent. At New Orleans, ranked No. 4, the increase was 39.5 percent. And at Beaumont, Texas's port, ranked sixth-busiest in the nation, freight tonnage increased a staggering 160 percent.

Business at other ports is booming as well, and the boom is expected to continue. O'Neill writes that Long Beach, Los Angeles, and New York/New Jersey predict that by 2020 they will be handling three times as much containerized freight as they did in 2000. Indeed, O'Neill summarizes nicely the mounting pressure that escalating waterborne freight places upon the entire intermodal infrastructure: "Congestion," he writes, "is a big problem not only at ports, but also on the rail and road networks that connect them to communities, consumers, and businesses." Instead of gateways, ports will become "economic bottlenecks if they fail to double or triple their cargo-handling capacity over the next 20 years." (*Blueprint: Ideas for a New Century*, 46)

Two fully-loaded container ships waiting to unload their cargo at the Port of Long Beach.

As we are forced to confront the myriad quality-of-life problems resulting from congestion and declining mobility, it's time to flip the coin over and remind ourselves just how much an efficient transportation infrastructure system means to our national health and personal well-being. Among the many statistics that might be adduced to make this point, here are a few of the salient ones:

More than 75 percent of the nation's products, with a value of more than $5 trillion annually, reach their destinations via our nationwide system of highways.

Investment in highway construction and improvements accounted for one-quarter of the post-World War II increase in American business productivity, and the return on investments in highways exceeded the return on private sector investments such as manufacturing plants and equipment.

Designing, building, maintaining and managing the nation's transportation infrastructure generates more than $200 billion annually in U.S. economic activity and sustains jobs for 2.2 million Americans.

Alan Pisarski also argues cogently for the national investment in transportation: "Putting money into the transportation system is an enabling act that rationalizes our investments in homes, offices, farms, factories, and vehicles. The costs of a first-class system are less than the costs of wasted fuel, lost time, and squandered opportunities inevitable under an inefficient system. An effective system will pay for itself in saved lives, saved time, saved tempers, increased social and economic opportunity, and an improved environment." (*Blueprint: Ideas for a New Century*, 20)

In making its case for the reauthorization of the federal surface transportation programs, ARTBA rightly calls the nation's highway system "the warehouse of American businesses." Just as our transportation infrastructure has driven the phenomenal growth of our economy in the 20th century, so it must in the 21st. The speed and efficiency promised by "e-commerce" will mean little without an efficient surface transportation system. For the continued expansion of American business, our transportation infrastructure must

likewise expand—by highway, by rail, by air, by water. It is the spreading arterial system on which the great body of our economy depends for its health and growth.

But even beyond what it means for business, think what our transportation infrastructure means to the daily lives of the American people . . . what it means to be able to go to school, to go to a concert or movie, to go to the grocery store; what it means to be able to send and receive gifts and other goods by air; what it means to have the whole world's produce brought to our shores and to our doors by water, rail, and roadway.

This all-inclusive mobility has become one of our nation's defining freedoms—guaranteed not by the Bill of Rights, but by a renewed national commitment to the world's best transportation infrastructure.

UPS jets on tarmac at night.

No traffic problems for these business women on commuter rail platform.

chapter nine

INTRODUCTION

By Alan E. Pisarski, author of *Commuting In America*

At the midpoint of the last century, Albert Einstein was asked for his perspective about what the remainder of the century might offer. He said that many really exciting concepts had been developed in the late 19th and early 20th centuries—radio, television, atomic energy, the automobile, air travel, rocket science and atomic energy—and that the rest of the century and beyond would focus on doing those things better than they had been done—perfecting them. I would say that the wonder shifted from being able to do these things at all to doing them well; and seeing what we mean by "doing them well" evolve. One thing we have learned is the power of "networked systems." This was a concept not well appreciated until recently. In 1997 the federal government considered the concept of networked systems in its classifications of industries—including power transmission, communications, water and sanitary systems and transportation. Recent analysis of the power of public transportation investment has shown that it is the network that generates great benefits. The tragic events of 2001 have reestablished in our minds the fundamental need for reliability and redundancy in our networks. Doing things well today means doing them with economic efficiency; doing them with recognition of their economic and environmental and social implications. We expect so much more of our transportation systems today and we will demand so much more of them in the future. The demands of the society for a responsive transportation network are greater than ever. But our skills are greater, our resources are greater, our understanding of human needs and purposes are greater as well. We cannot fail.

*"At the center of all these issues is the burden of time pressures
that most Americans feel . . . in a world that places increasing value on time,
even the same levels of travel time from one period to the next will be less tolerable."*
—*Alan Pisarski,* Commuting in America II

America's Transportation

*W*hat *We Know: A Look at the Numbers* ～ What we know for sure is that the future will bring us more people and more goods. The use of every mode of transportation will increase, and this increase will most certainly be led by the personal automobile. We also know that, to meet increasing demand, our transportation infrastructure must expand. As we have seen throughout the 20th century, the transportation construction industry stands ready to meet this challenge. Having also seen the immense dividends paid by investment in transportation, it is clear that additional investment will be necessary to keep America moving into the 21st century. ～ Dr. William Buechner, ARTBA's top economic analyst, predicts that we will have 246 million motor vehicles on America's highways by 2009, a 14 percent increase over 1999. By 2015, highway travel is expected to increase a staggering 40 percent. ～ Looking two decades out, the U.S. Census Bureau predicts that, from 2000 to 2020, total U.S. population will rise by 48.2 million people.

Network of the Future

That's about 12 million every five

years and a 17.6 percent total increase over a twenty

year span. Based on rates from 1980 to 1997, we can extrapolate

that, from 2000 to 2020, there will be 48 to 62 million more vehicles in the

U.S.—a 24 to 28 percent rise in vehicles over the 2000 total of 214 million. ∽

This growth *will* happen. Ideas for limiting future growth, says Brookings

Institution Senior Fellow Anthony Downs, are delusions. "Our challenge is to

accommodate growth, not prevent it.... Somehow, U.S. ground transportation systems must

expand their capacity to cope with this large increase in persons and households and goods." ∽

Robert E. Skinner, executive director of the Transportation Research Board, citing hard evidence that

automobile ownership is growing even faster than the population, speculates that we are reaching the

"saturation point." Households without automobiles declined from 20 percent to 8 percent between 1969

and 1995, he notes, while at the same time households with two or more motor vehicles increased

Terminal 4, the new international arrivals building at John F. Kennedy International Airport.

from 30 percent to 60 percent. All this happened while the average household declined from 3.14 to 2.65 persons between 1970 and 1995. Persons per vehicle declined from 2.6 in 1955 to 1.3 in 1995.

These trends are abetted, observes Skinner, by the perhaps surprising fact that automobile travel is getting cheaper all the time. "The inflation-adjusted state and federal gas tax at 1.6 cents per vehicle-mile in 1995 is near an all-time low. For comparison, it was 4.5 cents per vehicle-mile in 1965 and 2.5 cents in 1985. The average price of gasoline, including taxes and adjusted for inflation, is also near an all-time low, lower than it was just before the 1973 oil embargo."

"In a nutshell," Skinner declares, "Americans want it all—more mobility, more accessibility, more personal space, a better environment, and so on—and in the future, there will be more of us wanting these things.... Almost certainly, transport infrastructure will not keep pace with travel demand."

The pressures of increasing population upon highway travel will certainly spill over onto other modes of transportation. By 2010, the number of air passengers is expected to exceed one billion a year (up from 2001's 650 million), and

takeoffs and landings will increase 32 percent. The stress will be magnified by the fact that, of the nation's 548 commercial airports, 70 percent of passengers board at just 31 of them.

Our railway infrastructure, too, is feeling the pressure. According to the Progressive Policy Institute's Paul Weinstein Jr., despite Amtrak's problems, ridership is growing on its short regional routes and commuter rail routes in high-density areas, like the Northeast Corridor and areas along the West Coast. Weinstein reports that ridership in the Northeast Corridor grew 5 percent in 2000 and is up 8 percent in 2001. (*Blueprint: Ideas for a New Century*, 44).

As for local transit systems, the overall significance of this mode of transportation is sometimes obscured in the "congestion" debate. True, it is unlikely that transit will solve all of America's congestion problems, especially given predictions that the number of cars on the road will outpace any increase in transit ridership. Yet it's certain that congestion, particularly in crowded cities, would be much worse without transit, and it's also certain that transit ridership does continue to grow.

The American Public Transportation Association reports that over the past five years transit ridership has increased 21 percent, spurred by significant growth in Washington, D.C., and New York City. Of course, compared to automobile use, the number of transit users remain, and will remain, quite small, but the point is that the number of people using every mode of transportation continues to grow.

Some of the most remarkable growth in the transportation sector is taking place at our nation's ports. In 1999, some 100 U.S. ports handled nearly 800 million tons of cargo, valued at $600 billion. Yet even so, the port of New York/New Jersey handled 14 percent more cargo during the first quarter of 2001 than in the same period in 2000. More spectacularly, major ports on both coasts expect their volume of containerized cargo to triple between 2000 and 2020.

Ships themselves, on average, have doubled in size during the past 20 years, necessitating upgrades in port infrastructure. The Port Authority of New York and New Jersey, for example, plans to spend $1.4 billion over the next seven years to expand and improve its terminal facilities.

But what's happening at our ports is part of a much larger pattern: the transport of the ever-growing volume of freight—locally, nationally, and internationally—that constitutes the core of American business. Recently the Federal Highway Administration (FHWA) Office of Freight Management and Operations commissioned the Batelle Institute to produce a number of working papers that would analyze the trends and issues affecting freight transportation productivity in the United States. Some of their findings are offered in the paragraphs that follow.

On the global scene, international trade constitutes a growing share of the U.S. economy. Between 1960 and 1999, world merchandise trade (exports and imports) grew at an average annual rate of more than 10 percent, and this trend toward globalization has had a significant impact here at home. As a share of the gross domestic product, U.S. trade in goods and services has grown from 10.7 percent in 1970 to 26.9 percent in 1999.

However, if the U.S. is to take full advantage of the economic benefits of participation in the global economy, it must address the problem of infrastructure capacity shortfalls. According to Cambridge Systematics, the Batelle Team member that prepared the FHWA working paper on global trade, all transport modes that serve international trade are experiencing growing congestion at key hub facilities. Ports, as noted, must find ways to accommodate ever-larger ships (such as the so-called post-Panamax vessels that are too large to transit the Panama Canal), at a time when obtaining land for expansion is exceedingly difficult. Air traffic, already overloaded in many key international airports, faces mounting pressure to handle the import-export trend toward packages of higher value, smaller size and increasing time sensitivity.

On-dock rail facilities and access for double-stack intermodal services also must be expanded, the report continues, and the problem is exacerbated by the fact that surface access for trucking to port and air hubs typically depends upon congested freeways and major arteries. Finally, the surge in intra-continental trade that has resulted from NAFTA threatens to overwhelm the transport infrastructure along the north-south corridor, especially at border crossings.

Containers await shipping at terminal facility in New York.

Panamax Bulk Carrier built by Japan's Hitachi Zosen Corporation.

Meanwhile, locally and nationally, web-based retailing (e-commerce) is a key component of the emerging business trend from "push logistics" (inventory-based) to "pull logistics" (replenishment-based). In simplest terms, this is a shift from manufacturing to keep inventories full to manufacturing to fill individual orders, and its impact on the shipment of goods is, of course, enormous. According to Reebie Associates, which prepared the FHWA paper "Economy," while e-commerce offers consumers a means to purchase more in less time, it imposes a heavy burden on the transportation system to ensure on-time delivery.

Indeed, "just-in-time" delivery is increasingly considered a component of service, as we see in such industries as home grocery delivery, Internet catalog shopping, and Dell Com-puters' to-your-door service—to name just a few. As manufacturers lure customers with the promise of speedy delivery, the demand for freight transport via airplane and truck intensifies, and as inventories shrink, the link between production and transportation must become stronger and more reliable. Ultimately it is up to the carrier to make the delivery, and, thanks to the pressures of competition, each day we see an increasing range of shipment priorities and levels of available service: today, next morning, next afternoon, two-day, etc.

But carriers know that "just-in-time," with its menu of delivery options, means more deliveries, primarily over regional highway networks, and that their ability to perform is finally dependent upon the condition and efficiency of those networks. As this report concludes, "A major concern [of freight carriers] is the planning and financing of basic investment in the transportation infrastructure. The creation of public/private partnerships for major freight projects must be viewed as a challenge, since the major freight carriers and shippers historically have had little interaction with the state, regional and local agencies responsible for public transportation policy and funding."

Clearly, then, the stress upon our nation's transportation infrastructure can only increase. More people, more goods, more congestion on highways and at air and rail terminals—and, of course, less time. Alan Pisarski, in *Commuting in America II*, astutely observes that "at the center of all these issues is the burden of time pressures that most Americans feel. It is time pressures, particularly on women, that increase personal vehicle use, trip chaining, and many of the other patterns we have examined." (100) Moreover, he adds, "in a world that places increasing value on time, even the same levels of travel time from one period to the next will be less tolerable." (44)

Perhaps it's time to throw our hands up in surrender, or to admit, with Anthony Downs, that congestion "is a problem without a solution—at least no solution the American people will accept."

On the other hand, maybe it's simply time to go back to the drawing board; time to call once more upon the ingenuity and dedication of the engineers, contractors, manufacturers, operators and workers who make up the transportation construction industry; time to support their efforts with a vote of public confidence and federal support. It's a formula that has proven highly successful for a hundred years.

Assuming, then, that our problems are not insoluble, let's take a hopeful glance at what the world of transportation might look like two decades into the new century.

A Federal Express driver delivers an assortment of overnight packages.

What We Can Dream: A Glimpse at the Year 2020

The personal automobile, with the comfort, convenience, and flexibility American consumers have come to rely on, will remain the preferred mode of transportation as far into the future as anyone can reliably see. But cars will travel within a transportation infrastructure that has been enhanced by significantly added capacity and by the development of ITS technology.

Thanks to the ongoing Intelligent Transportation Systems (ITS) revolution, automobiles and the roads they travel on will become "smarter" than they are today. The computer-driven message boards already suspended above

commercial-free news, music, and entertainment.

The next step in automotive intelligence, writes Glen Hiemstra in *The Futurist,* will be built-in sensors that will view the road in front, in back, and on the side, and produce a 3-D view for the operator. These sensors will establish safe zones around the car and apply accelerator and brakes to maintain the zone.

While they're getting smarter, of course, cars will also become more fuel efficient. Responding to government pressure, auto manufacturers have already made huge gains in fuel economy and clean-engine technology, but cleaner-burning internal combustion engines may themselves soon retreat before the advance of new power-source technology—best exemplified by the increasing presence on our highways of so-called "hybrid" cars. These cars, like

In-vehicle information systems may be part of transportation's future.

Honda Insight, one of the gasoline-electric hybrids currently on the market.

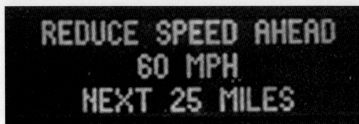

our freeways, giving drivers real-time information about traffic conditions ahead, are a promising first step. Soon to come will be advanced route guidance information and, perhaps most welcome, radically improved incident management—as police and emergency personnel will know immediately the exact location of any traffic problem.

While no one predicts that ITS will have solved our congestion problems by 2020, the in-vehicle component of ITS known as telematics—built-in electronics and wireless communication technologies—promises to make the time we spend in our cars more productive and enjoyable. In addition to traffic-related real-time information, roadside assistance, and turn-by-turn navigation, telematics systems already on the market are offering Internet and e-mail access, along with satellite radio with multiple channels of

Toyota's Prius and Honda's Insight, overcome the power-storage limits of electric (battery-powered) engines by combining them with a small, fuel-efficient internal combustion engine—thus making them "hybrid."

Are consumers ready? Yes, answers futurist Hiemstra: "The Batelle Institute forecasts that such hybrid cars will be a top-10 consumer product by about 2006." But the hybrids are only a small step in the power-source evolution. "Following the hybrids by a few years," says Hiemstra, "will be cars driven by fuel cells . . . [which] generate electricity through a chemical reaction between hydrogen and oxygen. They produce little waste except water. Practical fuel-cell technology could be available to the market as early as 2010."

The ultimate ramifications of this technology are far-reaching, says Hiemstra: "With this development, the internal combustion engine can be completely eliminated, and we could see a shift in the auto paradigm more sudden than most imagine. Hybrids and fuel-cell cars will extend the auto-based lifestyle far into the future."

Of course, for transportation planners, this "shift in the auto paradigm" has reverberations that Hiemstra fails to consider. Since highway user fees—primarily the gasoline and diesel excise taxes—are

Engineer at University of Minnesota's Human Factors Research Laboratory cruises "virtual highway" through Tofte, Minnesota.

One artist's vision of the automobile of the future.

overwhelmingly the most important source of funding for the nation's road improvements, the demise of the internal combustion engine will necessitate an equally sudden and radical shift in the "funding paradigm."

However, other visionaries are already looking beyond fuel-cell power and exploring the concept of "roadway power." Christopher Finch reports on experiments currently underway testing the viability of "roadway-powered electric vehicles," equipped with conventional battery storage systems for neighborhood driving, but "supplied with power pickups mounted beneath the car for travel over highways provided with a power source." In this case, electric cables buried just beneath the road surface "would transfer their power inductively across the air gap (about one inch) between the road and the pickup device. No mechanical contact is necessary, so the car can maneuver freely on the highway just as if equipped with a conventional engine." (381)

Of course, the cleanest power in the world will have no mitigating effect on congestion. In fact, the availability of "clean" automobiles might induce environmentally sensitive consumers to aban-don transit, thus making our roadways more crowded than ever and the problems facing our highway infrastructure all the more urgent.

True, the route-guidance and incident-management components of ITS have some potential for alleviating congestion. But for real impact upon highway mobility—and for an interesting new challenge for the transportation construction industry—we must take yet a further glance into the co-evolution of the car and the highway. Francis D. Reynolds calls it "dualmode."

Writing in *The Futurist*, Reynolds agrees with other trend-watchers that the use of private cars can only increase, but, he declares, "by using vehicles in two distinct modes we will avoid most of the transportation problems we now have." In Mode One, cars will be driven in the normal manner on

A high percentage of the private dualmode commuting cars will be small, two-passenger vehicles. Small cars tend to be more dangerous on the highways, where they have to compete with heavier cars and trucks. Size won't affect safety on the guideways because all vehicles will always travel at identical speed."
—*Francis D. Reynolds,* THE FUTURIST

These are computer graphics simulations of an automated highway system. Vehicles and roadways would be equipped with computers to provide controlled—and safer—driving.

Electric-powered CyberTran, being developed in Alameda County, California, is computer controlled and travels up to 150 mph.

W*ith vehicles moving at 60 mph, a single guideway lane could carry the traffic of 12 highway lanes; at 200 mph, one guideway lane would be equivalent to 40 highway lanes."*
—*Francis D. Reynolds,* THE FUTURIST

neighborhood or city streets, but in Mode Two, for longer trips, "they will travel automatically on high-speed dedicated guideways, while their drivers use the time for reading, talking, or almost anything but driving." Unlike automated highway systems that typify today's ITS thinking, says Reynolds, with dualmode most of the artificial intelligence is built into the guideway network, not the cars.

The applicability of the system is universal, according to Reynolds. The guideways will carry almost all of the categories of vehicles now used on streets and highways, and in complete safety. "A high percentage of the private dualmode commuting cars will be small, two-passenger vehicles. Small cars tend to be more dangerous on the highways, where they have to compete with heavier cars and trucks. Size won't affect safety on the guideways because all

vehicles will always travel at identical speed."

The answer to congestion lies in the fact that the guideways will operate nonstop at full speed, day and night, "just as our highways would if they weren't jammed with traffic." Reynolds proposes "a constant system speed of 60 mph in and around cities, and a constant 200 mph on the guideways between cities." Vehicles entering the guideway will be synchronized, enabling them to travel perhaps as close as one foot apart, "like boxes on a conveyor belt. . . . This speed offers enormous system capacity," notes Reynolds. "With vehicles moving at 60 mph, a single guideway lane could carry the traffic of 12 highway lanes; at 200 mph, one guideway lane would be equivalent to 40 highway lanes."

Here's how it works: To enter the guideway system, we drive into an entry stop, shut off the motor, and punch the number of the desired guideway exit into a keypad on the dashboard. That exit number will tell the navigation computer where to send us and enable the billing computer to charge us for that particular trip.

In "street mode," says Reynolds, cars will be battery-electric or fuel-cell powered. "But since these cars will be using guideway power for most of their travel, batteries or hydrogen tanks that are now inadequate for highway use will be more than adequate for dualmode in its limited street use."

The guideways themselves, of course, will be powered by electricity, which can come from any energy source—including sun and wind. "We will have many options in generating electricity," Reynolds writes, "but with internal-combustion automobiles we will have very few."

Dualmode will ease pressure on other modes of transportation as well, says Reynolds—particularly aviation. "The 200-mph guideways will make 'driving' faster than flying for thousand-mile trips," he predicts, "eliminating the hassles of airport transit, parking, late or canceled flights, and baggage and security checking. . . . In addition, air freight traffic will be greatly reduced by driverless freight containers on the high-speed guideways."

If Reynolds' vision becomes a reality, of course, it will be yet another remarkable testament to the ingenuity of the transportation construction industry—and yet another example of the public/private partnership that has driven the development of America's transportation infrastructure all along. Reynolds approximates that a dualmode system for the United States will cost $20 to $50 billion per mile of guideway—hundreds of billions for the whole nationwide system. Perhaps the guideway system will ultimately pay for itself in user fees analogous to today's gas tax, but that will be after the fact of its planning, design and construction. What about in the meantime?

Reynolds sums up the case for dualmode with the observation that "the only way to get high capacity and safe high speeds, keep the advantages of private cars, and solve most of our transportation-related environmental problems is to design and build a revolutionary transportation system using 21st-century technology. We will do this because there is no other satisfactory choice." Perhaps we will. But it will require a renewed commitment of the public will to meet our nation's transportation needs.

But what about inside our cities, where the guideways presumably will not penetrate? Given continued growth in population and car ownership, won't our city centers become more congested than ever?

If you're traveling alone—as is the case with the overwhelming majority of commuters—imagine this. Exiting the guideway, you pull into the nearby "transport station," where, sheltered in a vast garage and plugged into an electrical outlet, your Segway sits waiting for you. You're now entering Dean Kamen's vision of transportation's future.

Kamen tells *Time* magazine that his much-anticipated invention, the Segway, will be to the car what the car was to the horse and buggy. "Cars are great for going long distances, but it makes no sense at all for people in cities to use a 4,000-lb. piece of metal to haul their 150-lb. asses around town."

To bring the future into the present, Kamen, already a prolific and prosperous inventor of medical equipment, developed his revolutionary vehicle at a cost of more than $100 million. *Time's* John Heilemann describes it as "a complex bundle of hardware and software that mimics the human body's ability to maintain its balance. Not only does it have no brakes, it also has no engine, no throttle, no gearshift and no steering wheel. And it can carry the average rider for a full day, nonstop, on only five cents' worth of electricity."

Dean Kamen leans his body to initiate a turn on his Segway.

"Cities need cars like fish need bicycles," says Kamen, though he acknowledges that his invention is not likely to have any impact upon the already developed cities of today. Rather he hopes that the planned cities—or communities—of the future will be structured to accommodate his dream: with a mass-transit system encircling the city and the central area reserved for pedestrians, bicyclists, and, of course, Segway riders. (*Time*, Dec. 10, 2001)

In the meantime, the occasional Segway that ventures out into today's metropolis will probably find itself equally unwelcome on the sidewalks, where it travels too fast, and on the streets, where it travels too slowly.

Riders test the one-person, battery powered scooter in a New York city park.

Will the Segway prove useful? This postal carrier in Tampa may have the answer.

If our streets and highways will be taking on a new look by 2020, so will our skies. In his new book *Free Flight*, author James Fallows foretells the arrival of a nationwide air taxi fleet that will overthrow the tyranny of the current hub-and-spoke system. According to Fallows, the new system will make use of thousands of underused local airports for direct flights between small and medium-sized cities, along with the latest developments in small airplanes: the four-passenger, single-piston SR series from the Cirrus Design Corporation with a price tag of under $200,000, for example, or the five-passenger jet from Eclipse Corporation that will cruise at 41,000 feet, fly more than 400 miles per hour, and sell for less than $1 million. Fallows believes that when the "new age" arrives, many people will be able to travel the way the very wealthy do now—in comfort, avoiding congested hubs, leaving from close to where they live and landing close to where they want to go, at their own schedule and at a cost no higher than coach fares.

But if more, smaller aircraft flying to more, smaller airports doesn't seem quite revolutionary enough, Glen Hiemstra suggests that personal transportation may be ascending from the ground to the sky. "The Jetsons may have had it right all along," he writes. "Small, private flying vehicles suitable for short-haul personal transportation are in the experimental stages."

Hiemstra reports that Moller International is developing one such vehicle, the Skycar—a flying machine that features four redundant engines and complete computer control of flight. The redundant engines assure safety, and computer control means that anybody can operate one. Current estimates are that the Skycar will get 15 miles to a gallon of fuel, fly at 350 miles per hour, and sell for about the cost of a luxury automobile.

So perhaps the coming decades will offer innovative solutions to our looming crisis in mobility. Perhaps the transportation revolution we have traced throughout the 20th century will continue unabated.

But while we wait for dualmode guideways and personal aircraft, there is a lot of work to be done.

Skycar on exhibit at Big Boys Toys exhibition in Sydney, Australia, 2001.

In the Meantime: Getting from Here to There

It seems likely that successfully bridging the gap between today and the high-tech transportation future of tomorrow will require two coordinated courses of action: making better use of the transportation infrastructure in place, and building to meet demand.

As we have seen, the central idea behind ITS is making optimum use of the infrastructure already in place. In addition to the ideas already discussed, one already seeing limited application is turning High-Occupancy Vehicle (or HOV) lanes, which have been widely underused, into HOT (High-Occupancy and Toll) lanes, which would be open not only to high-occupancy vehicles but also single-occupant vehicles willing to pay for the privilege. Another option is building self-financed "truck only" lanes, particularly in congested corridors.

As noted in the preceding chapter, restructuring landing fees in such a way as to better spread air traffic across the span of hours offers another way to get the most out of the infrastructure we have. Similarly, says Hugh O'Neill, our ports will be able to handle expected increases in cargo volume not so much by becoming bigger but by be-

Amtrak's Acela, outside its factory in Barre, Vermont

Truck traffic is heavy on Chicago's Dan Ryan Expressway. Truck-only lanes could alleviate congestion.

coming more efficient. O'Neill suggests that our ports can improve on this score "by investing in modern equipment and information technology, by extending their operating hours, and by curbing the practice of letting shippers leave cargo awaiting pickup at the port for several days without charge." *(Blueprint: Ideas for a New Century*, 47)

As for our passenger rail lines, Skinner suggests that we can make good use of the infrastructure in place if we improve it toward "high-speed" capability by "incrementally upgrading existing passenger rail lines with electrification, selected alignment improvements, track improvements, and 'tilt trains,' such as Amtrak's new

Acela, which will tilt as it goes around curves to increase passenger comfort by lessening the effects of centrifugal force." Building new high-speed rail systems is another option in some corridors.

Alaska Representative Don Young, chairman of the Transportation Committee in the U.S. House, introduced a $71 billion high-speed rail bill in 2001—more than quintupling the size of the measure that had been on the table previously. During the Clinton administration, the federal Department of Transportation designated 10 regional rail corridors eligible for federal planning assistance including the Northeast Corridor; routes connecting major cities in Florida, Texas, the Midwest and the Northwest; and a north-south route in California.

Building additional capacity must certainly be a part of the solution to the worsening problems of congestion and mobility decline. Few serious transportation planners can doubt the critical need for additional airport runways, more and better transit services, enhanced intermodal facilities, and, yes, more lane-miles of highway.

No doubt some of the infrastructure work we see in the near future—adding additional expressway lanes in and around crowded cities, for example—will be dismissed by critics as quick-fix thinking and "business as usual." But other ideas are on the table, ideas that indeed look to the future: double-decked highways for dramatic increases in capacity, truck-only lanes to facilitate freight movement, elevated railways, and, yes, underground highways. Of course, in this litigious age, the boldest ideas—those most in need of vision, commitment, and investment—are also most likely to attract controversy.

There will be obstacles, debates both financial and philosophical, but it's worth reiter-

ating that investment in transportation has historically yielded rich returns for the nation's economy. The expansion of America's transportation infrastructure must proceed, and the work of the transportation construction industry will be increasingly spectacular and visionary, exemplifying the most remarkable 21st century innovation and technology.

While scientific genius is at work to bring us the technological advances touted by the futurists, the transportation construction industry—with the support of the American people and its government at all levels—will be hard at work on the nation's transportation infrastructure.

We are not there yet. But engineers, designers, contractors and construction workers across the land will tell you, "We are working on it."

Boston's four rapid transit lines are dispatched at this state-of-the-art Operation's Control Center located in the heart of Boston's Financial District.

Books

Warren Belasco. *Americans on the Road* (Cambridge: MIT, 1979).

Robert A. Caro. *The Power Broker* (New York: Knopf, 1974).

Dan Cupper. *The Pennsylvania Turnpike: A History* (Applied Arts Publishers, 1990).

Jonathan Daniels. *The Man of Independence* (New York: Lippincott, 1950).

Richard O. Davis, ed. *The Age of Asphalt* (New York: Lippincott, 1975).

Dwight D. Eisenhower. *At Ease: Stories I Tell to Friends* (New York: Doubleday, 1967).

James Fallows. *Free Flight: From Airline Hell to a New Age of Travel* (New York: Public Affairs, 2001).

Federal Highway Administration. *America's Highways, 1776-1996* (U.S. Department of Transportation: Washington, D.C., 1976).

Robert H. Ferrell, ed. *Dear Bess: The Letters from Harry to Bess Truman* (New York: Norton, 1983).

Christopher Finch. *Highways to Heaven: The AUTO Biography of America* (New York: HarperCollins, 1992).

Patrick J. Fitzpatrick. *Natural Disasters: Hurricanes* (Santa Barbara, Calif.: ABC-CLIO, 1999).

James Flink. *The Car Culture* (Cambridge: MIT, 1975).

Mark S. Foster. *From Streetcar to Superhighway: American City Planners and Urban Transportation, 1900-1940* (Philadelphia: Temple Univ. Press, 1981).

Deborah Gordon. *Steering a New Course: Transportation, Energy, and the Environment* (Washington, D.C.: Island Press, 1991).

Wesley S. Griswold. *A Work of Giants* (New York: McGraw-Hill, 1962).

Lawrence Halprin. *Freeways* (Reinhold, 1966).

James T. Jenkins, Jr. *The Story of Roads* (American Road Builder: Washington, D.C., 1967).

John Jackles. *The Tourist* (Univ. of Nebraska, 1985).

Thomas Jefferson. *The Portable Thomas Jefferson*. Merrill D. Peterson, ed. (New York: Viking, 1975).

Tom Lewis. *Divided Highways: Building the Interstate Highways, Transforming American Life* (New York: Penguin Books, 1999).

Stan Luxenberg. *Roadside Empires: How the Chains Franchised America* (New York: Viking, 1985).

Philip P. Mason. *A History of American Roads* (Rand McNally, 1967).

———. *The League of American Wheelmen and the Good Roads Movement, 1880-1905* (Univ. of Michigan, 1958).

David McCullough. *The Great Bridge* (New York: Simon & Schuster, 1972).

Clay McShane. *Down the Asphalt Path: The Automobile and the American City* (New York: Columbia Univ. Press, 1994).

Richard Lawrence Miller. *Truman: The Rise to Power* (New York: McGraw-Hill, 1986).

Alan Pisarski. *Commuting in America II* (Eno Transportation Foundation, 1996).

John Rae. *The Road and the Car in American Life* (Cambridge: MIT, 1971).

David A. Remley. *Crooked Road: The Story of the Alaska Highway* (New York: McGraw-Hill, 1976).

Mark Rose. *Interstate: Express Highway Politics* (The Regents Press of Kansas, 1979).

Eric Saul and Don Denevi. *The Great San Francisco Earthquake and Fire, 1906* (Millbrae, Calif.: Celestial Arts, 1981).

The States and the Interstates: Research on the Planning, Design and Construction of the Interstate and Defense Highway System (American Association of State Highway and Transportation Officials: Washington, D.C., 1991).

Gordon Thomas and Max Morgan Witts. *The San Francisco Earthquake* (New York: Stein and Day, 1971).

Margaret Truman. *Harry S. Truman* (New York: William Morrow, 1973).

Harry S. Truman. *The Autobiography of Harry S. Truman*, edited by Robert H. Ferrell (Colorado Associated Univ. Press, 1980).

John Edward Weems. *A Weekend in September* (College Station: Texas A&M Univ. Press, 1957).

Charles W. Wixom. *Pictorial History of Roadbuilding* (Washington, D.C.: American Road Builders' Association, 1975).

Periodicals and Presentations

Robert D. Atkinson, "The New Politics of Mobility," *Blueprint: Ideas for a New Century*, Sep./Oct. 2001.

William Bunch, "Robert Moses: His Legacy," *Newsday*, Dec. 4, 1988, p. 18ff.

Anthony Downs, "The Future of U.S. Ground Transportation from 2000 to 2020," testimony presented to the Subcommittee on Highways and Transit of the House Committee on Transportation and Infrastructure, March 21, 2001.

The Editors, "Keep America Moving," *Blueprint: Ideas for a New Century*, Sep./Oct. 2001.

Larry Flynn, "He Changed Our Lives," *Roads & Bridges*, Nov. 1999, p. 9ff.

William Fulton and Paul Shigley, "The Longest Day," *Governing*, Dec. 2001.

Glen Hiemstra, "Driving in 2020: Commuting Meets Computing," *The Futurist*, Sep./Oct. 2001.

Nancy Honssinger and Gina Hilton, "Country Kitchens," *Bittersweet*, Spring 1975.

Gary Hunter, "Preserving a Sense of Wildness," *Public Roads*, Spring 1995, p. 95ff.

Les Jackson, "Lean and Clean," *Blueprint: Ideas for a New Century*, Sep./Oct. 2001.

Tom Kuennen, Series of Articles Outlining ARTBA's 100-Year History in *Transportation Builder*, Jan.-Oct. 2001.

"Motor Car Triumphs in Crisis," *Motor Age*, May 3, 1906, p. 1ff.

Hugh O'Neill, "Gateways to Growth," *Blueprint: Ideas for a New Century*, Sep./Oct. 2001.

C. Kenneth Orksi, "Getting Smart Behind the Wheel," *Blueprint: Ideas for a New Century*, Sep./Oct. 2001.

Phil Patton and Barron Storey, "Agents of Change," *American Heritage*, Dec. 1994, p. 88ff.

Alan Pisarski, "Life in the Not-So-Fast Lane," *Blueprint: Ideas for a New Century*, Sep./Oct, 2001.

Francis D. Reynolds, "The Transportation System of the Future," *The Futurist*, Sep./Oct. 2001.

Dorothy Robyn, "Runway Gridlock," *Blueprint: Ideas for a New Century*, Sep./Oct. 2001.

San Francisco Chronicle, April 19, 1906.

Robert F. Skinner Jr., "Transportation in the 21st Century," paper presented to the Johns Hopkins Applied Physics Laboratory, June 9, 2000.

"The Superhighway Superman," *U.S. News and World Report*, Dec. 27, 1999, p. 51ff.

Walter L. Sutton Jr. and David Marks, "Highways and the New Wave of Economic Growth," *Public Roads*, Jul.-Aug. 1999.

Leonard Wallock, "The Myth of the Master Builder," *Journal of Urban History*, Aug. 1991, p. 339ff.

Richard F. Weingroff, "The Federal Highway Administration at 100," *Public Roads*, Fall 1993, p. 93ff.

Paul Weinstein Jr., "Putting Rail Back on Track," *Blueprint: Ideas for the New Century*, Sep./Oct. 2001.

Bill Wilson, "'Mr. Highways': A Legend Passes," *Roads & Bridges*, Nov. 1999, p. 26ff.

Conrad L. Wirth, "Mission 66," *American Road Builder*, Dec. 1959, p. 12ff.

Internet

American Life Histories: "Manuscripts from the Federal Writers' Project 1936-1940," Rose Wilder Lane:
http://memory.loc.gov/cgi/bin/query/D?wpa:2:./temp/~ammem_KUIv::

Doctor Gray:
http://memory.loc.gov/cgi/bin/query/D?wpa:13:./temp/~ammem_oIXO::

Indiana Historical Society: "The Hoosier Barnum: Carl G. Fisher"
http://www.indianahistory.org/heritage/fisher.html

The Learning Page: *Rise of Industrial America, 1876-1900*: "Rural Life in the Late 19th Century"
People Were Frugal:
http://lcweb2.loc.gov/ammem/ndlpedu/features/timeline/riseind/rural/frugal.html

Rosenberg Library: "Special Report on Galveston Hurricane," by Isaac M. Cline
http://www.rosenberg-library.org/gthc/clinereport.html

The Handbook of Texas Online: "LeTourneau, Robert Gilmour," by Ken Durham
(www.tsha.utexas.edu/handbook/online)